Challenging TripleCross Puzzles

Volume 1

Neil Aggarwal

For More Puzzles

Please visit http://3DMathPuzzles.com for more TripleCross, CrossFigure, and 3D puzzles!

Dedication

For my Dadhi, who's unconditional love is always in my heart.

Contents

Acknowledgments

This book could not be possible without a lot of help, especially from the following:

Jim Bumgardner (krazydad) whose excellent site http://krazydad.com provided many hours of enjoyment and inspired me to create puzzles of my own. He has also been very generous providing tips and support.

Yochanan Dvir whose cross-figure puzzles formed the reference material for the puzzles.

Jared McComb (bosuonparedo) who started me down the path to create the hexagonal puzzles.

Phil Bordelon (Sunfall) who suggested the name Triple Cross.

Dave Millar (Taco Dave) who suggested I create books.

All the users on the krazydad discord server who tested the puzzles and gave feedback.

Norma for her suggestions and editing help.

Sora Sagano for the Aileron font.

Charis SIL font Copyright (c) 1997-2022 SIL International (http://www.sil.org/) with Reserved Font Names "Charis" and "SIL". Licensed under the SIL Open Font License, Version 1.1, available at https://openfontlicense.org

Patrick Hand font Copyright (c) 2010-2012 Patrick Wagesreiter (mail@patrickwagesreiter.at), licensed under the SIL Open Font License, Version 1.1, available at https://openfontlicense.org

UXWing.com for the star icons.

jeonsango via Pixabay for the cover background image.

Canva for their tools which I used to design the cover image.

Instructions

This puzzle is like a crossword, but with numbers. Each digit occupies a hexagonal cell and can be a part of a "word" in the across, up, and down directions.

Rules:

1. "Words" may not start with a zero.
2. "Words" in the 'across' direction read from left to right.
3. "Words" in the 'up' direction read along the upward diagonal to the right.
4. "Words" in the 'down' direction read along the downward diagonal to the right.
5. There is one unique solution which satisfies all the clues given below the puzzle.

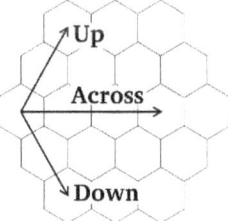

Some "words" may not have clues. They will be determined by the "words" which intersect them.

Solving Tips

I find it useful to keep track of the minimum and maximum values of each "word" in the puzzle. I repeatedly step through the clues to reduce the minimum and maximum values of each "word" as I apply the constraints given by each clue. The range between minimum and maximum can become narrower as you add constraints, but it can never get wider.

Keep in mind that all "words" must be integers. This has many implications, especially for multiplication and division. For example, doubling a "word" must give an even result. Also, you can only halve an even "word".

Keep in mind that all "words" must be positive. So, if you are subtracting 1 Across from 2 Down, it means 2 Down must be larger than 1 Across.

Keep in mind that all "words" must start with a non-zero digit. That means all the digits in labeled cells can't be zero. This especially helps when a labeled cell is in the middle or is the last digit of a "word". That will usually throw out some possibilities of the "word".

When any digit has limited possibilities, I keep track of its possibilities by writing them in the cell for that digit. I mark out the possibilities as I find clues which limit them until I get down to a single possibility. I especially focus on last digits of "words" since I can deduce them by looking at the last digits of the related words.

Every clue is a mathematical equation. Remember, equations can be rearranged. So, for example, if you have a clue where 1 Across is twice 2 Down, you can rearrange that to mean 2 Down is half of 1 Across. This is a simple example, but there are many algebraic manipulations you may perform on the clues.

Keep in mind that an even number plus an even number must result in an even number. Adding an even number to an odd number results in an odd number. And, adding two odd numbers results in an even number.

For subtraction, an even number minus an even number must result in an even number. Subtracting an even number from an odd number or an odd number from an even number results in an odd number. And, subtracting an odd number from an odd number results in an even number.

For multiplication, an even number multiplied by an even number must result in an even number. Multiplying an even number with an odd number results in an even number. And, multiplying two odd numbers results in an odd number.

For division, an even number divided by an odd number must result in an even number. Dividing an odd number by an odd number results in an odd number. And, an odd number divided by an even number will always result in a fractional number (which can eliminate a possibility if the result goes into a "word").

Since each "word" has a specific number of digits, you can throw out any result which gives you the wrong number of digits. So, for example, if you are doubling a two-digit number and looking for a two-digit result, the first number can only be in the range 10 to 49 and the result has to be an even number in the range 20 to 98.

If a "word" is prime, it can't end with zero (Since it would be divisible by ten), an even digit (It would be divisible by two), or five (It would be divisible by five). So, the last digit for any prime "word" can only be one, three, seven, or nine.

All square numbers end in zero, one, four, five, six, or nine since the last digit of the result is the same as squaring the last digit of the original number. So, if a "word" is square, the last digit can't be two, three, seven, or eight.

There are a limited number of square numbers which have a specific number of digits. This is especially true of a two-digit result since the only possibilities are 16,25,36,49,64, and 81. I usually don't apply this limitation to prime numbers because they have many more possibilities.

Puzzle 1

Difficulty: ★★☆☆☆

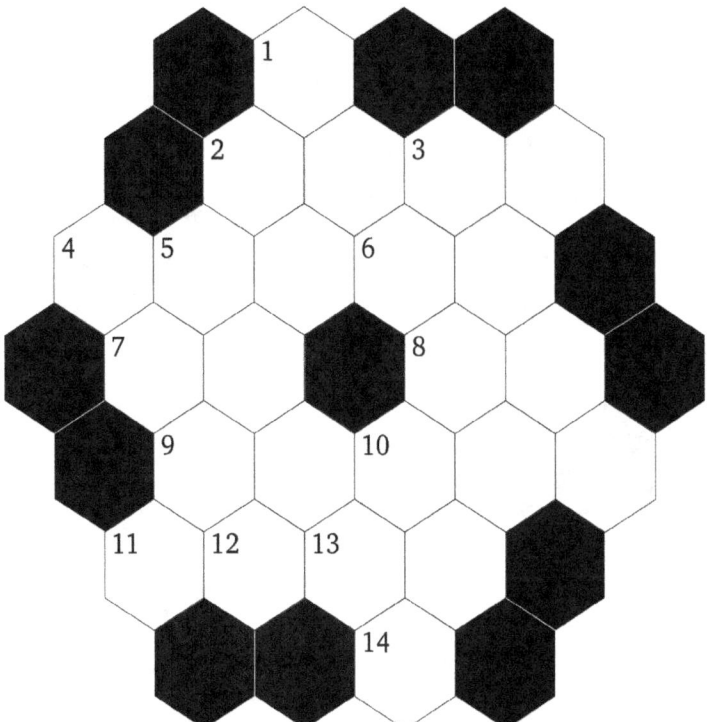

Across

2 Three times a prime number
4 Four times a prime number
7 8 across reversed
8 7 across reversed
9 Last three digits are the same as last three digits of 13 up
11 A square

Up

6 Consecutive digits in descending order
7 A square
12 Ten times a square
13 Three hundred thirty-one more than 5 down
14 Fifteen times a prime number

Down

1 Twice the result of 4 across minus 9 across
2 11 across divided by 10 down
3 5 down minus 4 down
4 A prime number
5 Eight times a prime number
10 A square

Puzzle 2

Difficulty: ★★☆☆☆

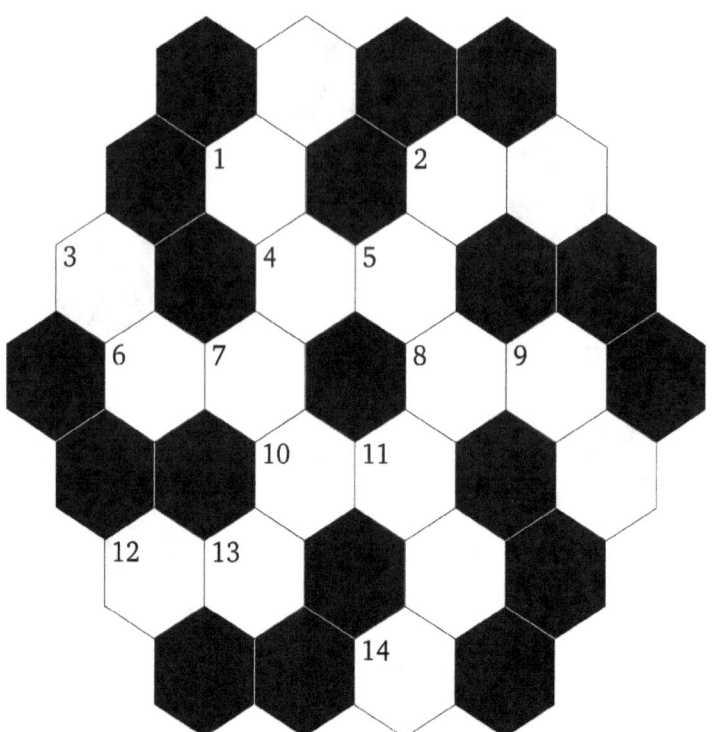

Across

2 A square
4 1 down minus 14 up
6 3 down plus 11 down
8 11 down plus 5 up
10 2 across plus 3 down
12 A square

Up

1 8 across minus 12 across
5 13 up minus 14 up
7 13 up plus 5 down
11 8 across reversed
13 Mean of 3 down and 7 up
14 1 up minus 2 across

Down

1 Three times 12 across
3 8 across minus 6 across
5 14 up plus half of 5 up
7 1 up plus 12 across
9 3 down plus 2 across
11 7 up minus half of 9 down

Puzzle 3

Difficulty: ★★☆☆☆

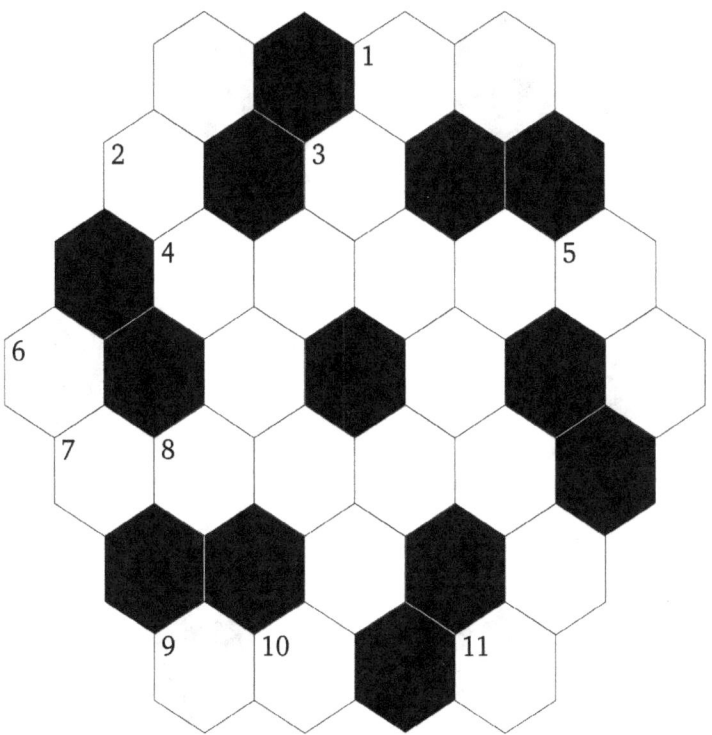

Across

1 Mean of 5 down and 11 up

4 Its digits total 1 across

7 2 down minus 2 up

9 11 up minus half of 5 down

Up

2 Twice a prime number

8 One thousand five hundred sixty-one more than 4 across

10 Five thousand one hundred thirty-five more than 7 across

11 Four times a prime number

Down

2 A prime number

3 6 down plus half of 10 up

5 Twice a square

6 A prime number

Puzzle 4

Difficulty: ★★☆☆☆

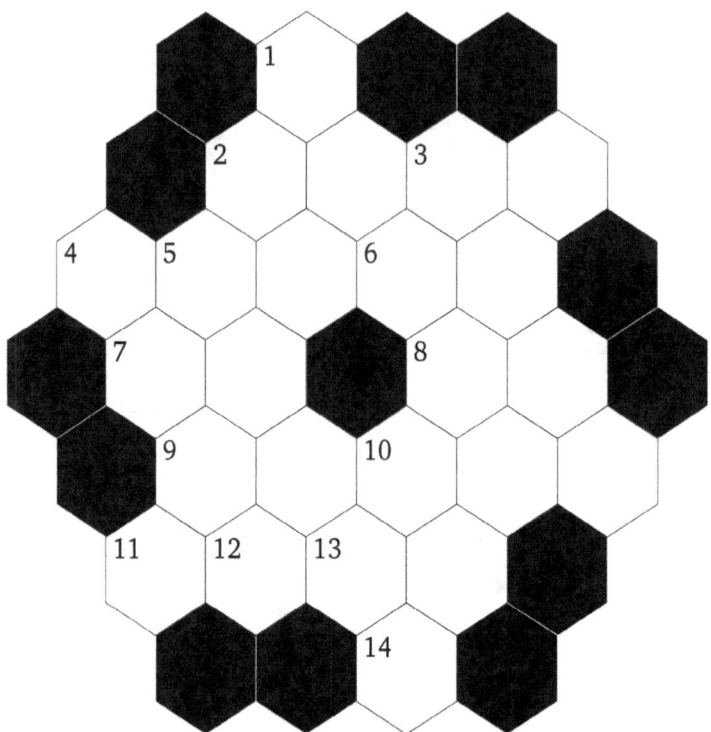

Across

2 Eleven times a square
4 Three thousand one hundred thirty-two more than 11 up
7 A square
8 A prime number
9 Ninety-four times a prime number
11 4 down minus 2 down

Up

6 Half of 9 across, then subtract 11 up
7 Sixteen times a prime number
11 Thirty-eight times a square
12 Same as 6 up
13 Thirty-seven times a prime number
14 13 up minus 5 down

Down

1 A prime number
2 Twice a prime number
3 Twice a prime number
4 Three times a square
5 Seventeen thousand nine hundred thirty-seven more than 1 down
10 All digits are the same

Puzzle 5

Difficulty: ★★☆☆☆

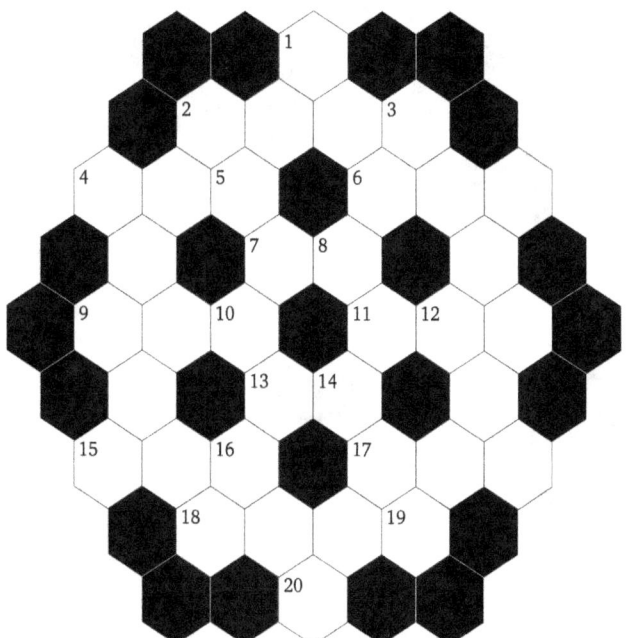

Across

2 Three times a prime number
4 Nine times a prime number
6 12 down minus 10 up
7 8 up minus 15 up
9 2 down minus 4 down
11 Nine times 14 up
13 A prime number
15 8 up plus half of 7 across
17 1 down minus 4 down
18 Half of 19 up, then subtract 17 across

Up

5 9 up minus 1 down
8 8 down plus 15 up
9 Twenty-eight times a prime number
10 2 down divided by fifty-one
12 Five times a prime number
14 Consecutive digits in descending order
15 Three-fourths of 18 up
18 16 down minus 9 across
19 Eight times a prime number
20 One hundred twenty-eight more than 12 up

Down

1 13 across plus 6 across
2 Mean of 20 up and 15 up
3 Six times a prime number
4 Three-fifths of 12 up
8 Twice the result of 18 up minus 14 down
9 A prime number
10 11 across divided by sixteen
12 A prime number
14 A prime number
16 Mean of 13 across and 12 up

Puzzle 6

Difficulty: ★★☆☆☆

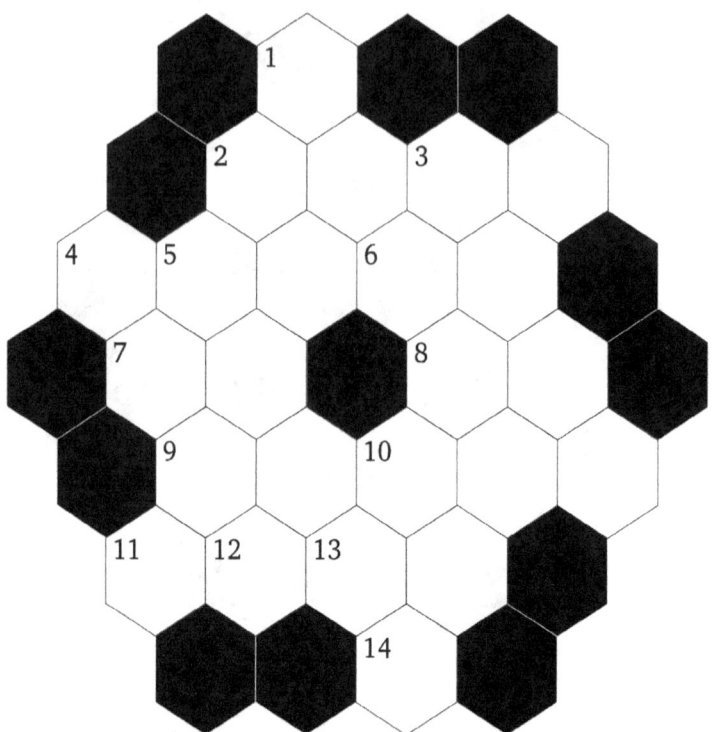

Across

2 Twice a prime number
4 Mean of 5 down and 2 across
7 Same as 12 up
8 Sum of digits in 7 up
9 Four times a prime number
11 4 down minus 7 across

Up

6 3 down divided by 10 down
7 Twenty-nine times a prime number
11 Five times a prime number
12 Five times a prime number
14 A prime number

Down

1 Twenty-seven times a prime number
2 Twelve times a prime number
3 Consecutive digits unordered
4 12 up plus 11 across
5 Last two digits are the same as 6 up
10 Consecutive digits in descending order

Puzzle 7

Difficulty: ★★☆☆☆

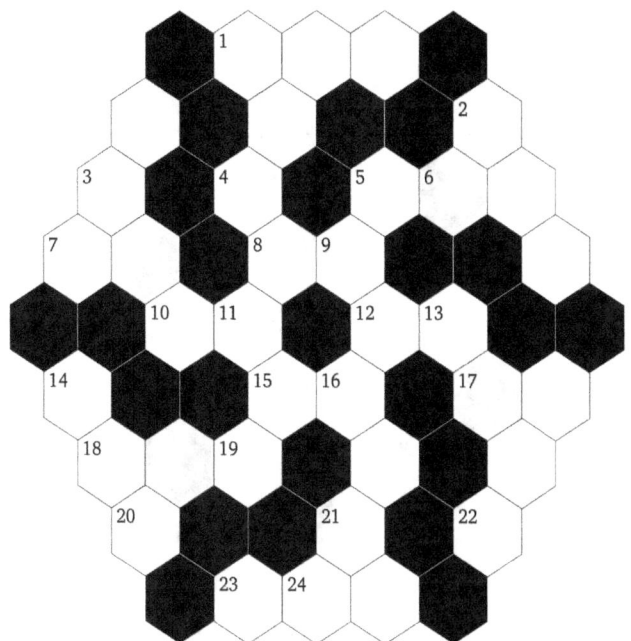

Across

1 Fifty-four times a prime number

5 Seven times a prime number

7 A square

8 Same as 9 up

10 4 up minus 9 down

12 6 up plus 17 across

15 6 up plus 8 across

17 10 across minus half of 11 up

18 A prime number

23 Three times a prime number

Up

4 21 down plus half of 9 up

6 15 across minus 9 up

7 Twice the result of 23 across minus 7 across

9 Mean of 20 up and 16 up

11 6 up plus 16 down

16 3 down minus 9 down

19 18 across minus 1 across

20 Three times a prime number

22 Mean of 5 across and 12 across

24 A prime number

Down

1 Mean of 4 up and 16 down

2 6 up times 17 across

3 19 up plus 7 across

4 Two-fifths of 19 up

9 15 across minus 4 down

11 Mean of 7 across and 21 down

13 Twice a prime number

14 Mean of 2 down and 8 across

16 Mean of 17 across and 3 down

21 A prime number

Puzzle 8

Difficulty: ★★☆☆☆

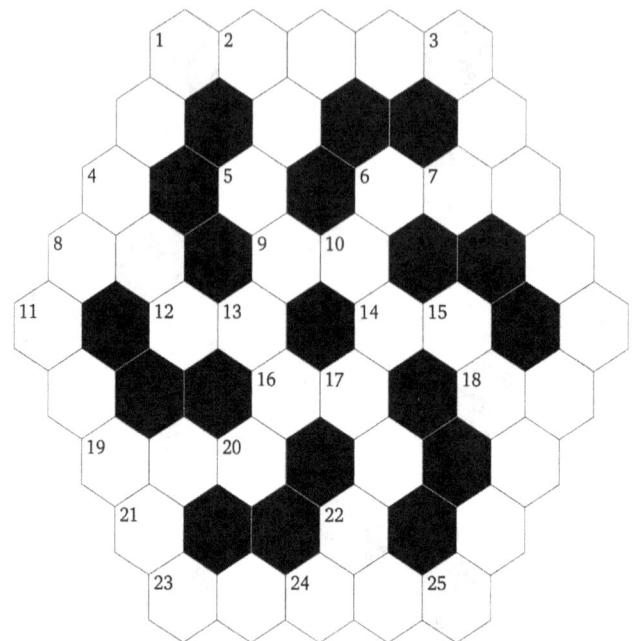

Across

1 Eleven thousand nine hundred fifty-eight less than 25 up
6 12 across plus 14 across
8 17 up divided by four
9 Mean of 2 down and 20 up
12 14 across minus 7 up
14 12 across plus half of 13 up
16 Four times 8 across
18 13 down minus half of 20 up
19 5 up minus 10 up
23 Mean of 1 across and 11 down

Up

5 19 across plus half of 9 across
7 11 up minus half of 25 up
10 12 across minus 2 down
11 Mean of 25 up and 13 up
13 Twice 7 up
17 Twice the result of 6 across minus 17 down
20 14 across minus 2 down
21 Mean of 10 up and 12 across
24 A prime number
25 Eighty-four times a prime number

Down

2 13 down divided by three
3 Twice a prime number
4 A square
5 6 across minus 7 up
10 Same as 22 down
11 Sixty-six times a prime number
13 17 down reversed
15 Five times a prime number
17 13 down reversed
22 Mean of 21 up and 8 across

Puzzle 9

Difficulty: ★★☆☆☆

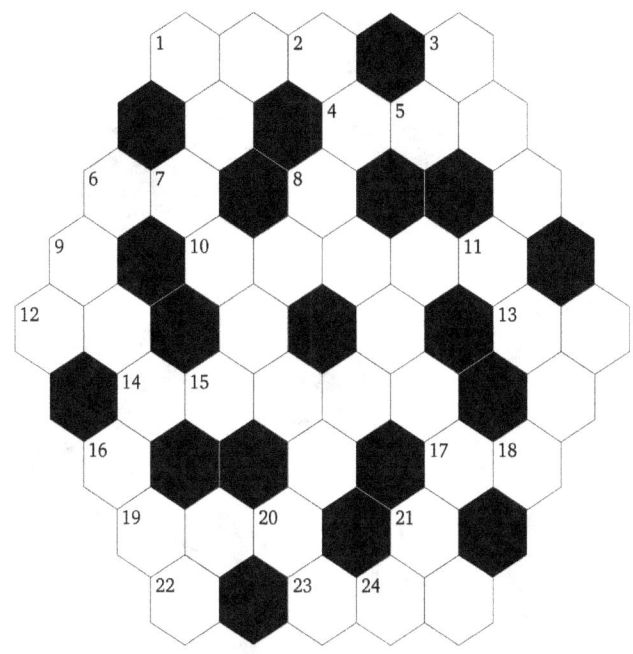

Across

1 21 down times 17 across
4 7 down divided by fifty-one
6 11 down divided by forty-one
10 12 up times 16 up
12 All digits are the same
13 14 across divided by 19 across
14 Nine thousand two hundred fifty-four more than 20 up
17 5 up minus 6 across
19 Four times a prime number
23 Its digits total 6 across

Up

5 14 across divided by 19 across
7 Mean of 2 down and 19 across
11 Four times 21 down
12 A square
15 Seventy-four times 3 down
16 Mean of 5 up and 21 down
18 Rearranged digits of 23 across
20 Three thousand two hundred forty-nine more than 10 across
22 Fifteen more than 13 across
24 Twenty-one more than 9 down

Down

1 A square
2 Seven times 6 across
3 Twice the result of 16 down minus 2 down
7 First two digits are the same as 20 down
8 Six times a prime number
9 One hundred twenty-four less than 1 across
11 20 up divided by 1 down
16 A prime number
20 Twice the result of 12 across minus 5 up
21 A prime number

Puzzle 10

Difficulty: ★★☆☆☆

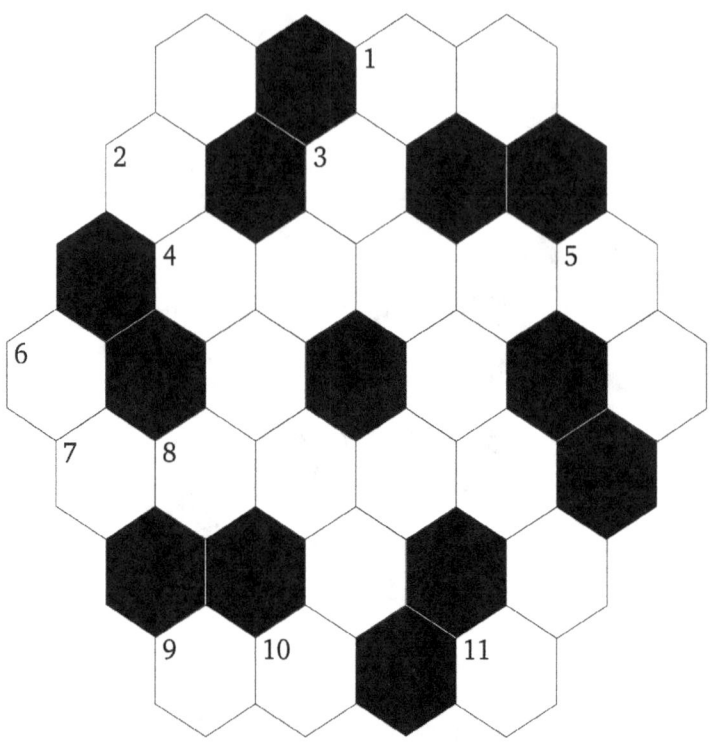

Across

1 Twice a square
4 Six thousand four hundred twenty-six more than 8 up
7 A prime number
9 Consecutive digits in ascending order

Up

2 A square
8 7 across plus 6 down
10 Twice the result of 2 down plus 9 across
11 1 across plus 2 up

Down

2 Fifty-three times a prime number
3 Rearranged digits of 10 up
5 Sum of digits in 3 down
6 Consecutive digits in ascending order

Puzzle 11

Difficulty: ★★☆☆☆

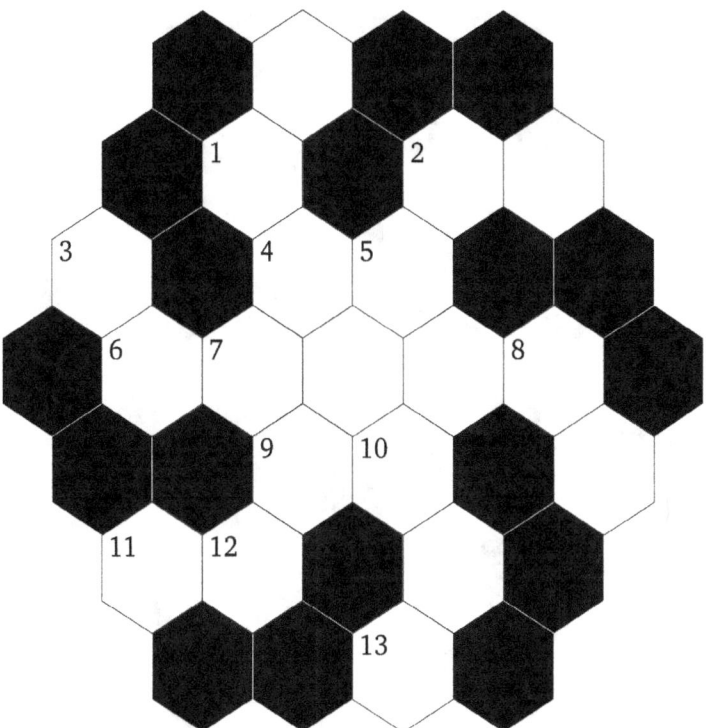

Across

2 5 down minus 13 up
4 7 up plus 1 up
6 Nineteen times a square
9 All digits are the same
11 Same as 2 across

Up

1 7 up plus 7 down
7 1 up minus 8 down
10 8 down plus half of 2 across
12 Twice the result of 6 across plus 1 up
13 5 down minus 11 across

Down

1 Thirty-seven times a prime number
3 Its digits total 7 up
5 A prime number
7 Mean of 7 up and 8 down
8 4 across minus 1 up

Puzzle 12

Difficulty: ★★☆☆☆

Across

3 Half of 11 up, then subtract 19 across
4 Twenty-two less than 18 up
7 19 across plus 13 up
10 14 across divided by fifty-seven
12 Mean of 2 down and 7 across
14 One thousand two hundred thirty-two less than 11 up
16 Seven times a prime number
19 Fourteen times a prime number
21 5 up minus half of 15 down
22 Three hundred seventy-one more than 19 across

Up

5 One thousand six hundred forty-six less than 8 down
6 Half of 24 up, then subtract 4 across
11 Twice a prime number
13 21 across divided by forty-five
17 Twice a prime number
18 Four times a prime number
19 Half of 2 down, then subtract 22 across
23 Twenty-seven times a prime number
24 Eight times a prime number
25 17 down minus half of 18 up

Down

1 A prime number
2 Forty-two times a prime number
4 Three hundred thirty-five more than 19 up
8 A prime number
9 Mean of 7 across and 14 across
11 13 up minus 17 up
13 20 down minus 4 down
15 Five hundred ninety-one less than 8 down
17 Thirteen times a prime number
20 Four hundred five less than 22 across

Puzzle 13

Difficulty: ★★☆☆☆

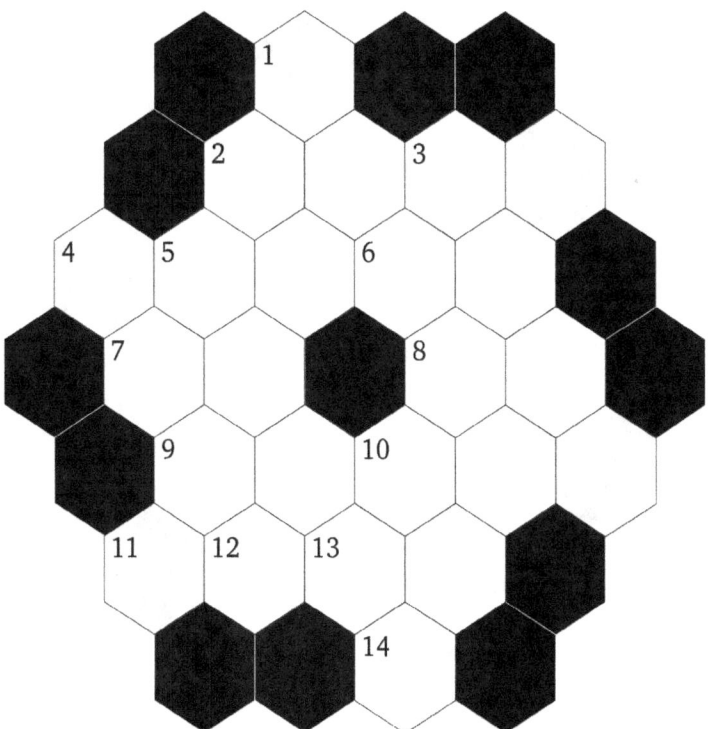

Across

2 Mean of 3 down and 6 up
4 Eight times a prime number
7 All digits are the same
8 Consecutive digits in ascending order
9 Seventeen times a prime number
11 Fourteen times a prime number

Up

6 Same as 12 up
7 Three times a prime number
11 A prime number
12 A prime number
13 Twenty-one thousand twenty-nine less than 9 across
14 Twelve times a prime number

Down

1 Mean of 4 across and 5 down
2 Half of 10 down
3 A prime number
4 Half of 14 up, then subtract 8 across
5 Twice a prime number
10 Sum of digits in 11 across

Puzzle 14

Difficulty: ★★☆☆☆

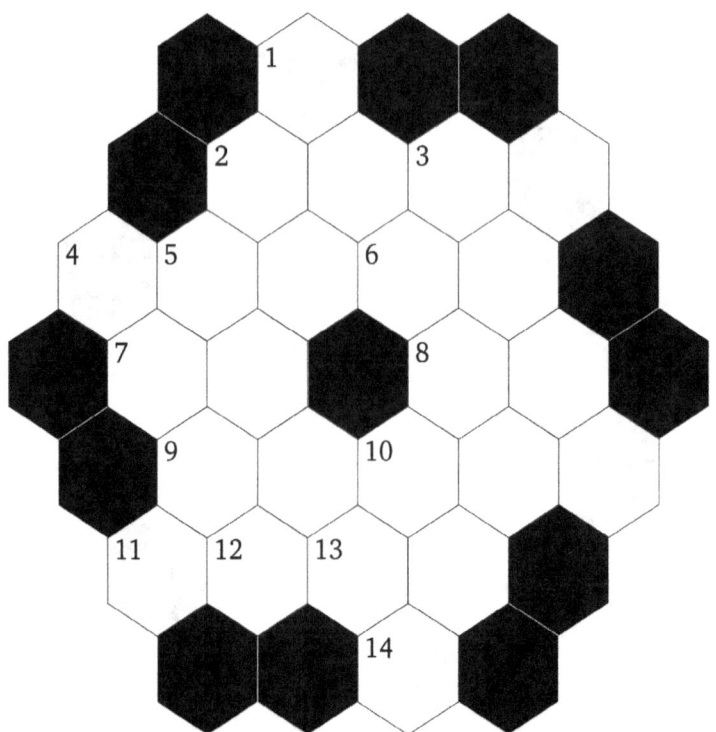

Across

2 Ten times a prime number
4 Twice the result of 13 up minus 6 up
7 A square
8 A square
9 A prime number
11 Twice a prime number

Up

6 A square
7 A prime number
11 Twenty-four times a prime number
12 11 up divided by 11 across
13 Four thousand one hundred nineteen less than 1 down
14 Sixteen times a prime number

Down

1 Thirteen times a prime number
2 Half of 14 up, then subtract 7 up
3 A cube
4 A prime number
5 Eight times a prime number
10 12 up plus 7 across

Puzzle 15

Difficulty: ★★☆☆☆

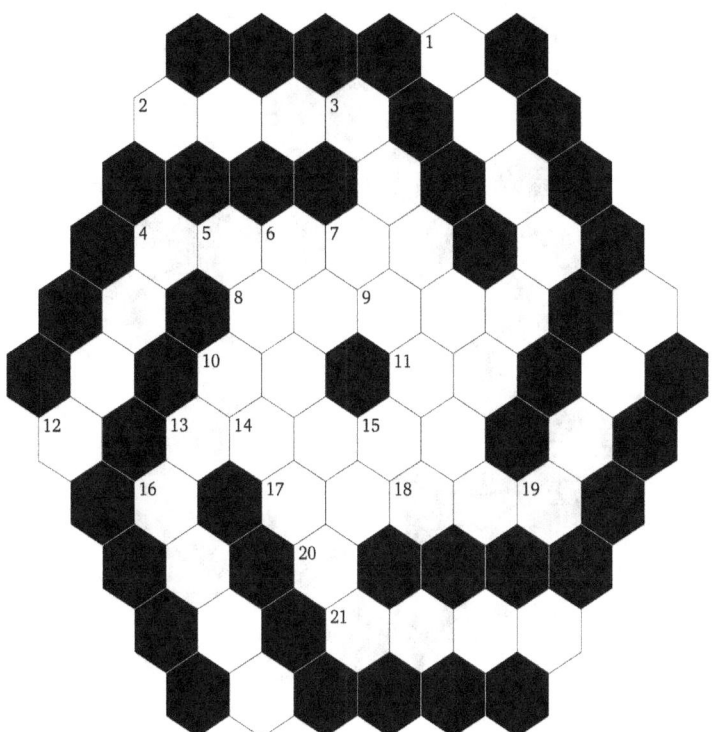

Across

2 Nine times a prime number

4 Seventy-four times a prime number

8 Five times a prime number

10 9 up divided by three

11 Consecutive digits in ascending order

13 Three thousand five hundred forty-six more than 18 up

17 Twice a prime number

21 Forty-nine times a prime number

Up

9 A square

12 A palindrome

14 12 up plus half of 10 down

16 Mean of 18 up and 19 up

17 Consecutive digits in descending order

18 Seven times a prime number

19 Fifty-seven times a prime number

20 Sixty-three times a prime number

Down

1 Fifty-seven times a prime number

3 Seven hundred forty-six more than 20 up

5 3 down minus 2 across

6 17 up plus 11 across

7 4 across minus 15 down

10 Thirty-eight times a prime number

15 Twice a prime number

16 1 down minus 21 across

Puzzle 16

Difficulty: ★★☆☆☆

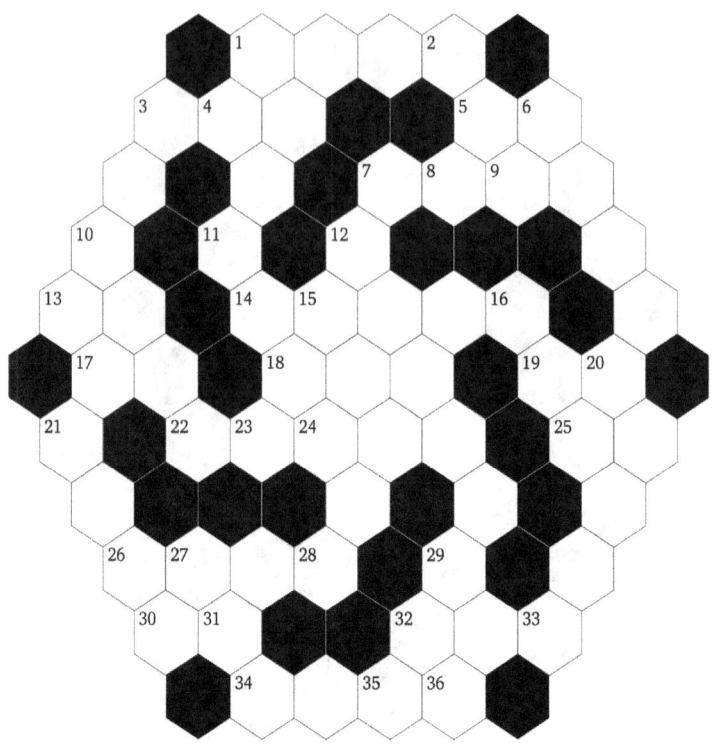

Across

1 21 down minus 20 down
3 2 down plus 8 up
5 21 up divided by 13 across
7 18 across times 19 across
13 8 up plus 19 across
14 Rearranged digits of 23 up
17 A square
18 13 across plus 29 down
19 Mean of 30 up and 13 down
22 Eleven times a prime number
25 21 down minus 1 across
26 A square
30 Same as 20 down
32 Mean of 24 up and 31 up
34 A square

Up

4 A square
8 20 down reversed
9 33 up divided by 20 down
11 Ninety times a prime number
13 Twenty-one times a prime number
21 Eighty-four times 31 up
23 First two digits are the same as 1 down
24 Nineteen times 4 up
25 A square
28 Six thousand three hundred eight more than 12 down
30 4 down minus 17 across
31 Half of 30 up
33 1 down times 30 across
35 Mean of 34 across and 32 down
36 17 across plus 13 down

Down

1 33 up divided by 25 across
2 Mean of 27 down and 32 down
4 A prime number
6 14 across minus 22 across
10 13 up plus 30 up
11 A prime number
12 Twenty-three times a prime number
13 1 down minus 4 up
15 29 down plus 4 down
16 2 down times 4 up
20 32 down minus 17 across
21 1 across plus half of 13 down
27 Four times a prime number
29 2 down minus 13 down
32 36 up minus 30 across

Puzzle 17

Difficulty: ★★☆☆☆

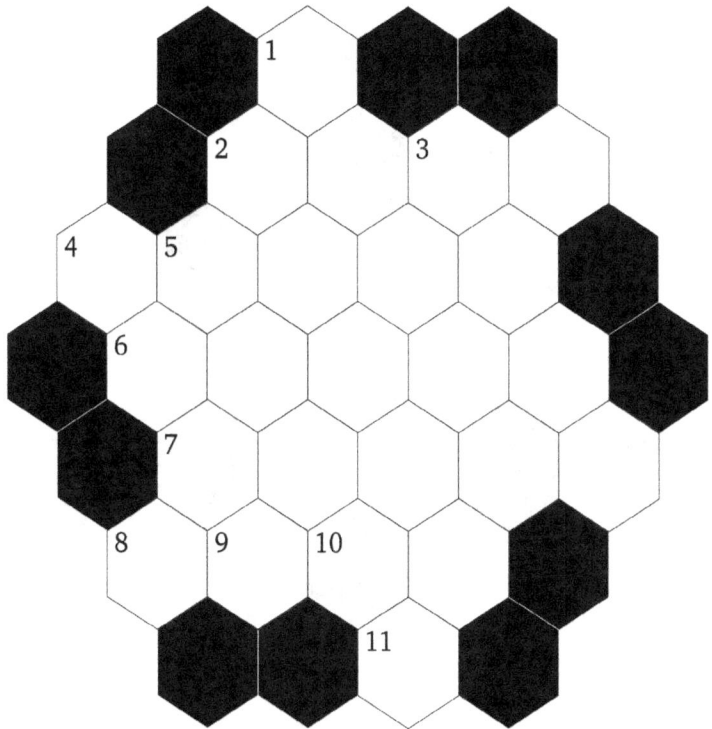

Across

2 Fifty-four times a prime number

4 Twelve times a prime number

6 Two hundred forty-two more than 10 up

7 11 up plus half of 2 across

8 Twenty-one times a prime number

Up

6 Twenty-six times a prime number

8 Twice a prime number

9 Thirty-nine times a prime number

10 Sixteen times a prime number

11 Nine hundred seven more than 8 across

Down

1 Half of 4 across, then subtract 6 up

2 A prime number

3 Thirty-nine times a prime number

4 Twice the result of 6 across minus 8 across

5 Sixty-six times a prime number

Puzzle 18

Difficulty: ★★☆☆☆

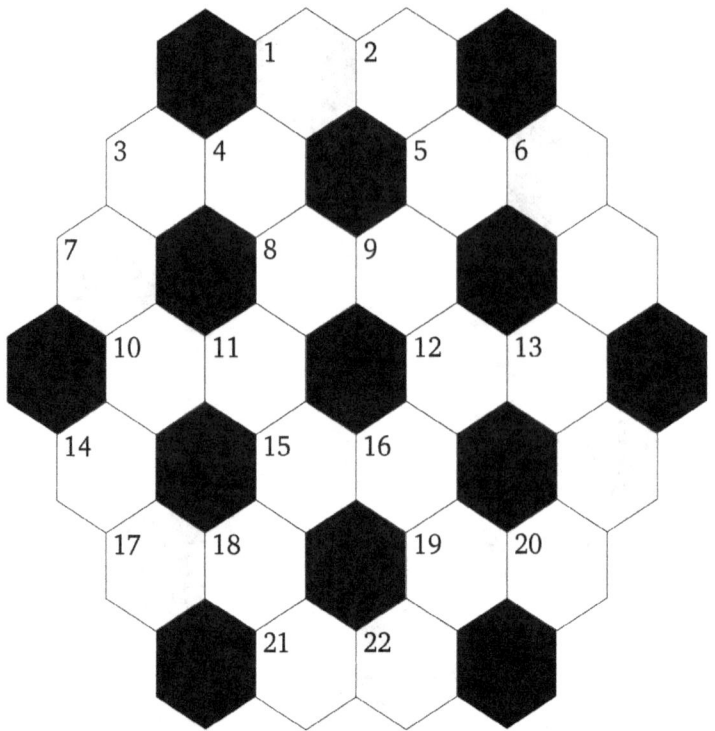

Across

1 12 across minus 18 up
3 16 down minus 9 up
5 Mean of 10 across and 1 across
8 Mean of 10 across and 18 down
10 Mean of 20 up and 18 up
12 Three times a prime number
15 16 down reversed
17 Mean of 9 down and 3 across
19 Half of 6 down, then subtract 22 up
21 Mean of 12 across and 4 up

Up

4 16 up minus 14 up
7 4 up minus 14 up
9 21 across minus 11 up
11 5 across minus 15 across
13 8 across minus 5 across
14 10 across minus 9 up
16 11 down plus 4 down
18 A square
20 Mean of 18 up and 8 across
22 Three-fourths of 11 up

Down

2 Twice the result of 13 down minus 11 down
4 9 down plus 15 across
6 Mean of 4 up and 8 across
7 14 down plus half of 17 across
9 A prime number
11 16 up minus 6 down
13 18 up minus 7 down
14 Twice 1 across
16 8 across plus half of 1 across
18 Twice the result of 18 up minus 3 across

Puzzle 19

Difficulty: ★★☆☆☆

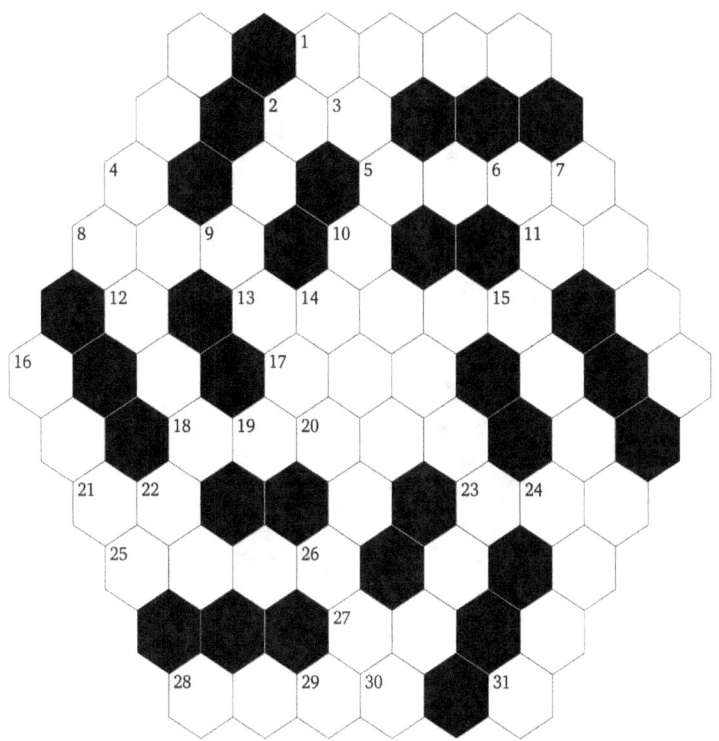

Across

1 Mean of 25 across and 12 up
2 15 up minus 12 up
5 Nine hundred eighty less than 30 up
8 6 down plus 2 across
11 12 up minus 29 up
13 A prime number
17 3 up plus 4 down
18 First two digits are the same as 24 down
21 Two-fifths of 22 down
23 1 down minus 11 across
25 Nine times a prime number
27 Five times a prime number
28 5 across plus 7 down

Up

3 Consecutive digits in descending order
8 24 up plus 8 down
9 Sixty-eight times 2 across
12 Mean of 24 down and 8 across
15 8 up minus 15 down
19 A prime number
20 Mean of 8 up and 22 down
24 30 up divided by 23 across
25 A prime number
26 Two thousand three hundred eighty-one more than 18 across
29 31 up divided by 2 across
30 One thousand ninety-six less than 28 across
31 Twenty-three times a prime number

Down

1 15 up plus 8 across
4 Three times 24 down
6 17 across minus 8 across
7 Twelve times a prime number
8 Last two digits are the same as 4 down
9 Six times a prime number
10 Mean of 26 up and 21 across
14 Half of 9 up, then subtract 8 down
15 Nineteen times 4 down
16 20 up plus half of 6 down
22 6 down reversed
24 15 down divided by fifty-seven
26 Thirteen times 21 across

Puzzle 20

Difficulty: ★★☆☆☆

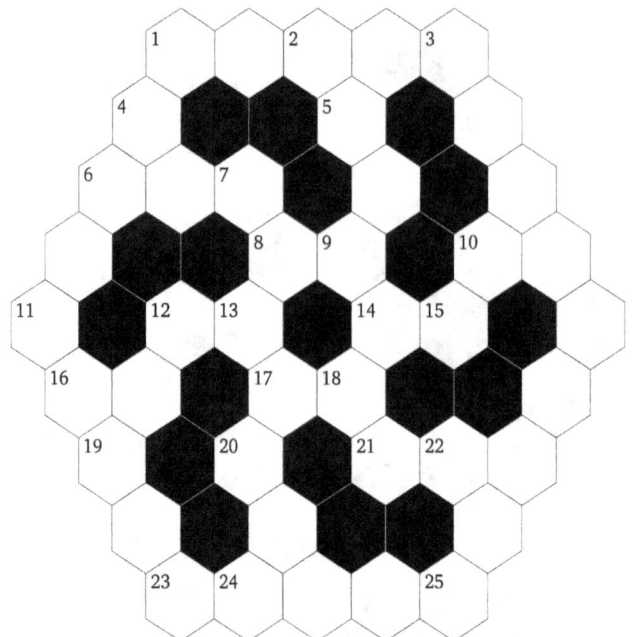

Across

1 Nineteen times a prime number
6 A square
8 13 down plus 22 down
10 15 up divided by 22 down
12 Mean of 4 down and 7 down
14 A square
16 4 down minus 17 across
17 2 down divided by 20 up
21 14 across plus 9 down
23 Three thousand five hundred twenty-six less than 11 down

Up

5 7 down minus 10 across
9 5 up minus 20 up
11 Forty-seven times a prime number
13 Mean of 21 across and 9 down
15 Twenty times 24 up
18 Mean of 18 down and 16 across
19 Three times 13 down
20 9 up minus half of 14 across
24 13 up minus 18 up
25 Sixty-two times a prime number

Down

2 11 up minus 23 across
3 1 across plus half of 25 up
4 18 up plus 10 across
7 Twice the result of 21 across minus 16 across
9 12 across minus 18 down
11 A palindrome
13 6 across minus 4 down
18 4 down minus 9 up
20 Mean of 19 up and 17 across
22 9 down minus 17 across

Puzzle 21

Difficulty: ★★☆☆☆

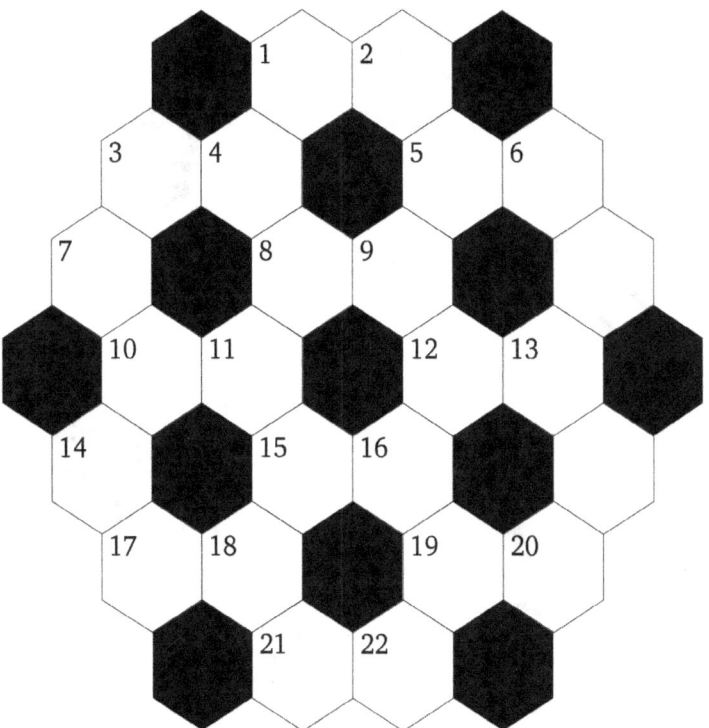

Across

1 Mean of 2 down and 10 across
3 1 across plus 4 down
5 7 down plus 8 across
8 11 up minus 18 down
10 4 up plus 8 across
12 11 up minus 16 down
15 22 up plus 13 up
17 Twice the result of 13 up plus 19 across
19 7 down minus 16 down
21 A square

Up

4 A square
7 5 across minus half of 6 down
9 8 across minus 9 down
11 2 down minus 4 up
13 20 up minus 14 up
14 16 up minus 20 up
16 20 up plus half of 13 down
18 21 across plus 12 across
20 Two-fifths of 5 across
22 Mean of 11 down and 21 across

Down

2 14 up plus 15 across
4 7 down minus 22 up
6 Mean of 9 up and 2 down
7 A square
9 13 down minus 4 up
11 2 down minus 9 down
13 16 up minus 13 up
14 7 down minus 10 across
16 Mean of 11 down and 21 across
18 Mean of 4 up and 1 across

Puzzle 22

Difficulty: ★★☆☆☆

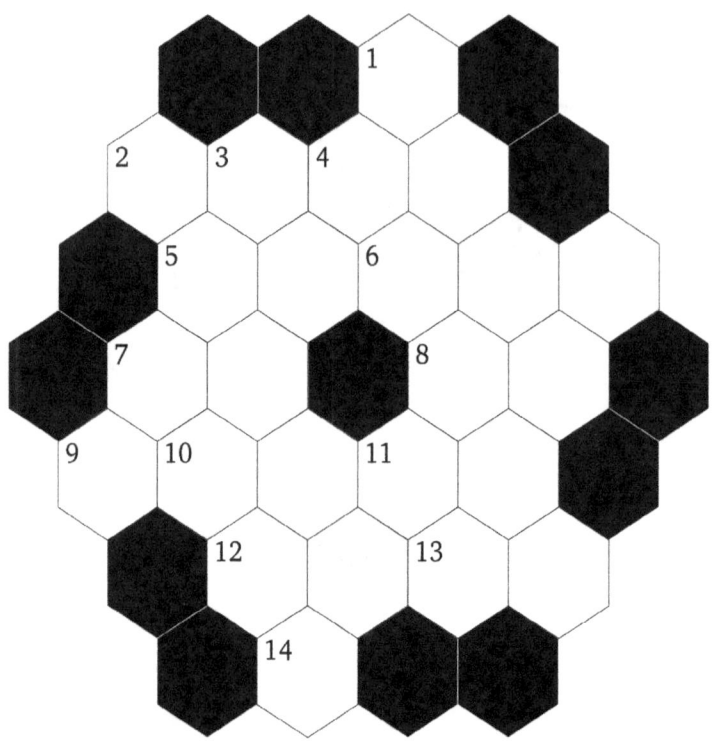

Across

2 A prime number
5 Fourteen times a prime number
7 Twice a prime number
8 Seven more than 6 up
9 Last two digits are the same as 11 down
12 A prime number

Up

6 A prime number
9 7 across times 6 up
10 Mean of 14 up and 13 up
12 Consecutive digits in descending order
13 Twice a prime number

Down

1 One thousand ninety-seven more than 2 across
2 Six thousand seven hundred sixteen more than 9 across
3 A square
4 5 across plus 12 up
7 Twice a prime number
11 Mean of 12 up and 3 down

Puzzle 23

Difficulty: ★★☆☆☆

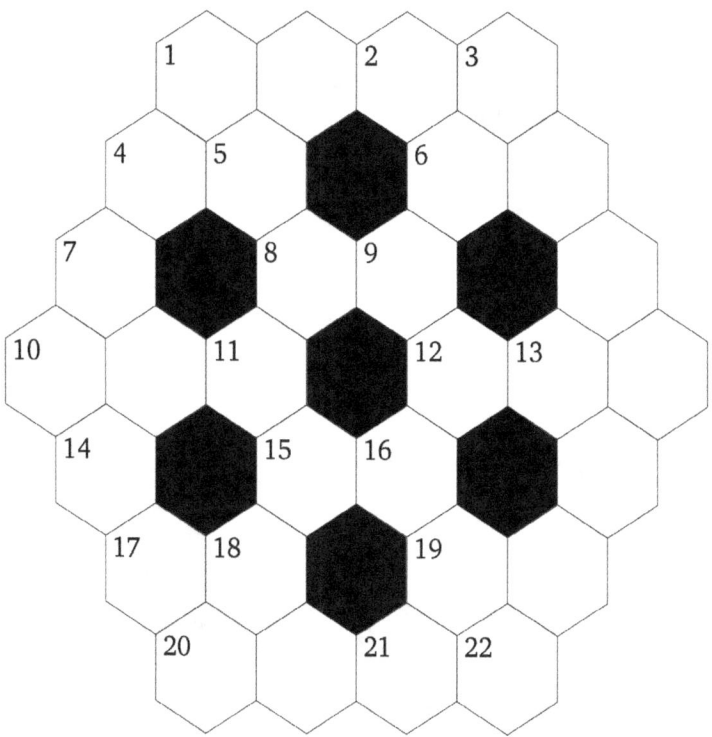

Across

1 16 down times 19 across
4 17 across minus 8 across
6 9 down plus 21 up
8 Mean of 9 down and 16 up
10 20 up plus half of 18 down
12 Twice a prime number
15 11 down reversed
17 18 down plus 11 down
19 14 up minus 11 up
20 Thirty-four more than 10 up

Up

5 6 across minus 18 down
9 Same as 12 across
10 Mean of 3 down and 11 up
11 8 across plus half of 7 down
13 14 up plus 19 across
14 Mean of 11 up and 13 up
16 17 across minus 19 across
20 Three times a prime number
21 20 across minus half of 3 down
22 Sixty-six times 11 down

Down

1 5 up times 18 down
2 21 up minus 7 down
3 Twelve times a prime number
7 6 across minus 14 up
9 14 up minus 2 down
10 13 down times 4 across
11 Mean of 13 up and 4 across
13 4 across plus 17 across
16 Twenty-seven times a prime number
18 14 up divided by six

Puzzle 24

Difficulty: ★★☆☆☆

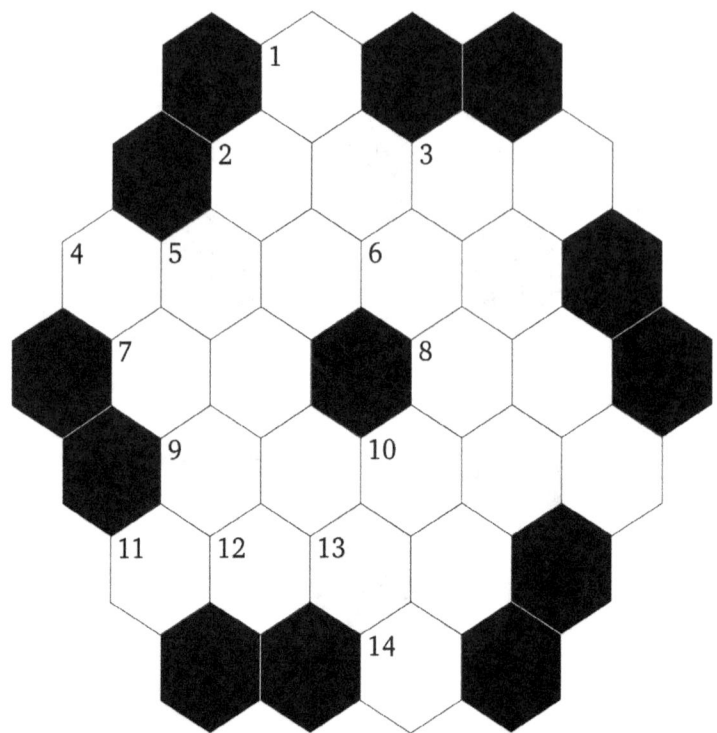

Across

2 Seven times a prime number

4 Fifteen thousand nine hundred seventy-seven more than 13 up

7 Mean of 6 up and 12 up

8 2 down reversed

9 Six times a prime number

11 Fourteen times a prime number

Up

6 All digits are the same

7 Fourteen times a prime number

11 One thousand eight hundred three less than 1 down

12 Twice a prime number

13 A prime number

14 A prime number

Down

1 Mean of 5 down and 7 up

2 8 across reversed

3 Mean of 11 across and 2 down

4 2 across plus half of 6 up

5 Twenty-six times a prime number

10 Twice a prime number

Puzzle 25

Difficulty: ★★☆☆☆

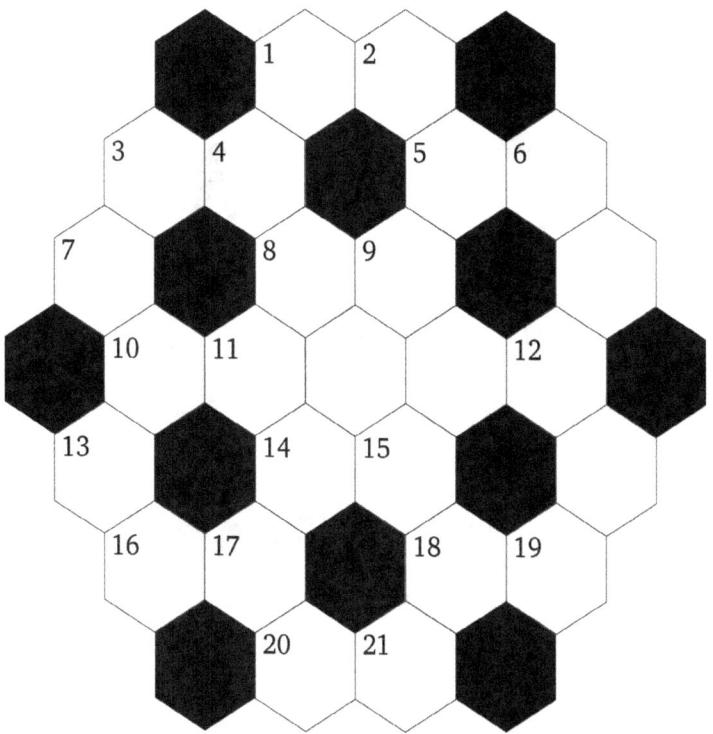

Across

1 Twice the result of 8 across plus 13 down
3 Twice the result of 6 down plus 7 up
5 7 up plus 12 down
8 Mean of 12 up and 7 down
10 Last two digits are the same as 13 down
14 Mean of 9 down and 18 across
16 Two-thirds of 20 across
18 21 up minus 19 up
20 18 across minus 9 down

Up

4 3 across minus 2 down
7 11 down minus 13 down
11 Mean of 9 down and 4 up
12 Half of 11 down, then subtract 16 across
13 1 across minus 9 down
15 14 across minus 13 up
17 Nine times a prime number
19 A square
21 11 down plus 11 up

Down

2 11 up minus 20 across
4 A prime number
6 4 up minus 20 across
7 Mean of 8 across and 2 down
9 Mean of 7 down and 21 up
11 2 down plus 15 up
12 A square
13 Mean of 17 down and 11 up
17 Mean of 5 across and 12 up

Puzzle 26

Difficulty: ★★☆☆☆

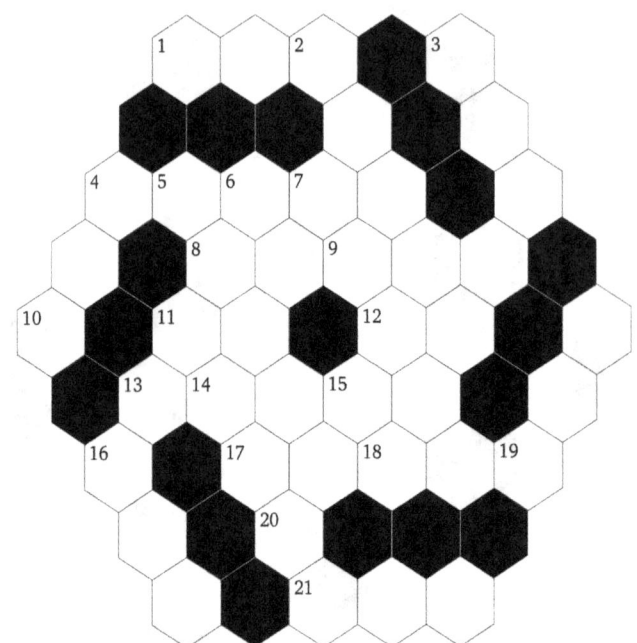

Across

1 Six times a prime number
4 Mean of 5 down and 10 up
8 Its digits total 9 up
11 Three-fourths of 15 down
12 Half of 1 across, then subtract 21 across
13 Eleven less than 17 across
17 Half of 18 up, then subtract 2 down
21 Eleven times a prime number

Up

9 11 across minus 17 up
10 Eight times 15 down
14 Nineteen times a prime number
16 11 down minus 16 down
17 3 down divided by sixteen
18 Twice a prime number
19 Four times a prime number
20 Thirty-four times a prime number

Down

2 Twelve times a prime number
3 Mean of 19 up and 15 down
5 Eight hundred fifteen more than 8 across
6 All digits are the same
7 A prime number
11 Two thousand two hundred ninety-one more than 4 across
15 Four times a prime number
16 Four times a prime number

Puzzle 27

Difficulty: ★★☆☆☆

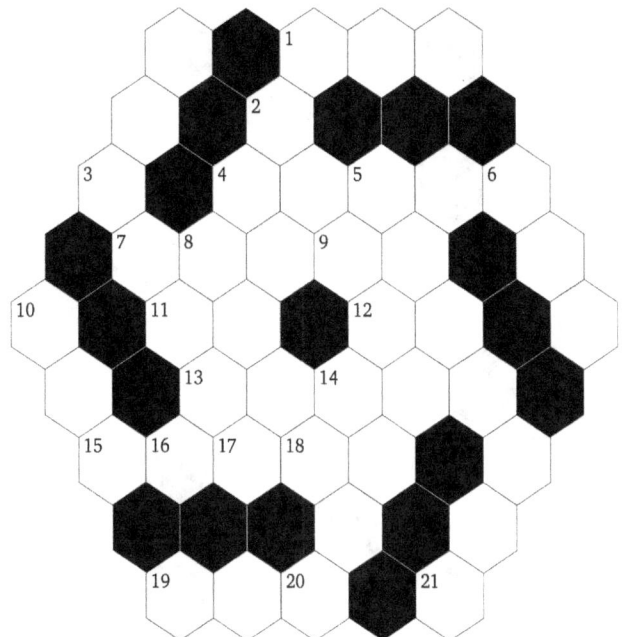

Across

1 Four times a prime number
4 Ten thousand one hundred fifty less than 2 down
7 Twice the result of 16 up minus 9 up
11 Five times a prime number
12 Three times a prime number
13 Ninety-three more than 20 up
15 Two thousand ninety-eight less than 8 down
19 Rearranged digits of 21 up

Up

3 Consecutive digits unordered
9 4 down reversed
11 3 up times 1 across
16 One thousand five hundred eighty-three more than 8 down
17 One more than 4 down
18 Fifteen times a square
20 Seven thousand three hundred seven more than 7 across
21 13 across divided by 17 up

Down

2 Seventeen times a prime number
3 Seven times a prime number
4 12 across plus 14 down
5 A prime number
6 Five times a prime number
8 Twice the result of 18 up minus 6 down
10 11 across plus 12 across
14 15 across divided by 19 across

Puzzle 28

Difficulty: ★★☆☆☆

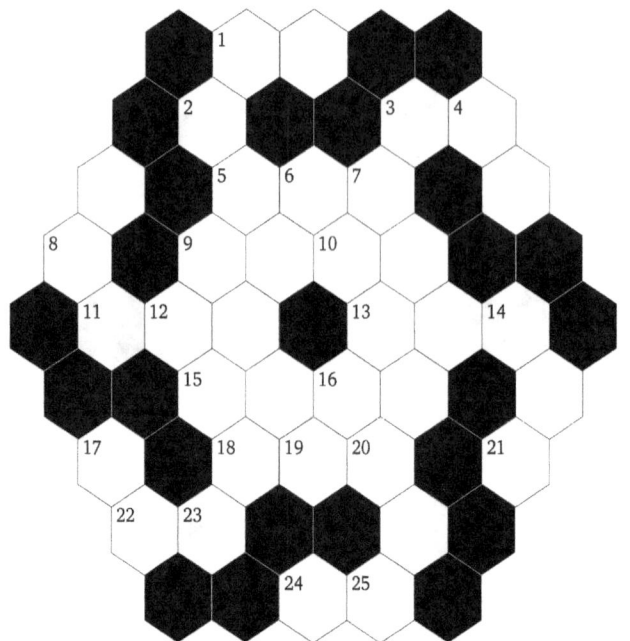

Across

1 15 across divided by 17 down
3 Mean of 24 across and 4 down
5 8 down plus half of 11 across
9 A prime number
11 Twice a prime number
13 14 down plus 24 across
15 Ninety-eight times a prime number
18 Same as 12 up
22 A prime number
24 A cube

Up

2 14 down reversed
8 Mean of 2 up and 25 up
10 9 down minus 2 down
12 23 up plus 14 down
15 One thousand three hundred eighty-two more than 6 down
19 Its digits total 21 up
20 Eleven times 22 across
21 24 across minus 4 down
23 A square
25 13 across minus half of 14 down

Down

2 A prime number
4 1 across minus 17 down
6 Twice a prime number
7 Three times a prime number
8 Mean of 2 up and 17 down
9 Seven times a prime number
12 Six times a prime number
14 18 across minus 23 up
16 12 down plus 8 up
17 24 across plus 8 down

Puzzle 29

Difficulty: ★★☆☆☆

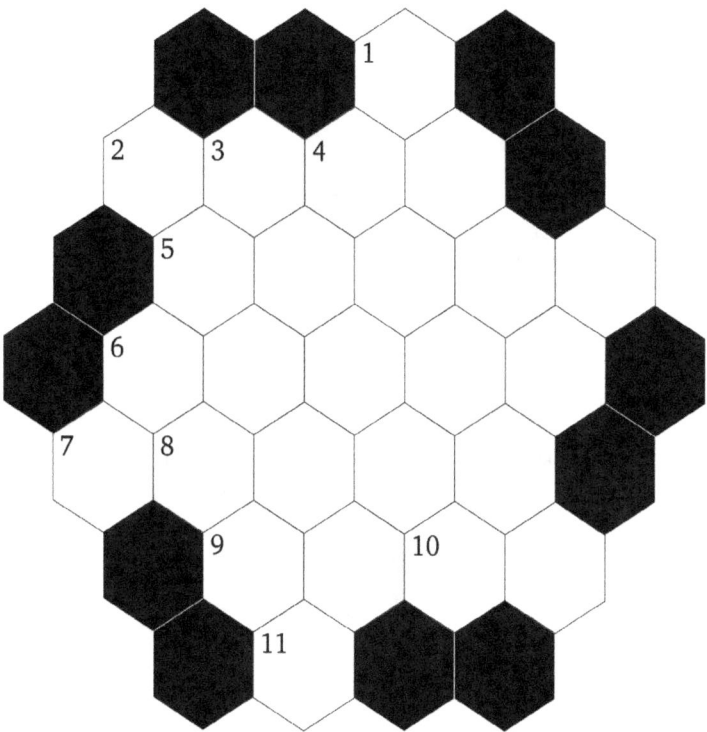

Across

2 Three times a prime number

5 Nine thousand eight hundred two more than 4 down

6 Eight thousand nine hundred seventy-nine less than 9 up

7 Four thousand nine hundred two more than 2 down

9 Thirty times a prime number

Up

7 8 up minus half of 9 across

8 Fourteen times a prime number

9 Fifty-nine times a prime number

10 Six times a prime number

11 1 down plus 2 across

Down

1 Consecutive digits unordered

2 Rearranged digits of 9 up

3 A prime number

4 Twice the result of 9 up minus 10 up

6 Thirty-one times a prime number

Puzzle 30

Difficulty: ★★☆☆☆

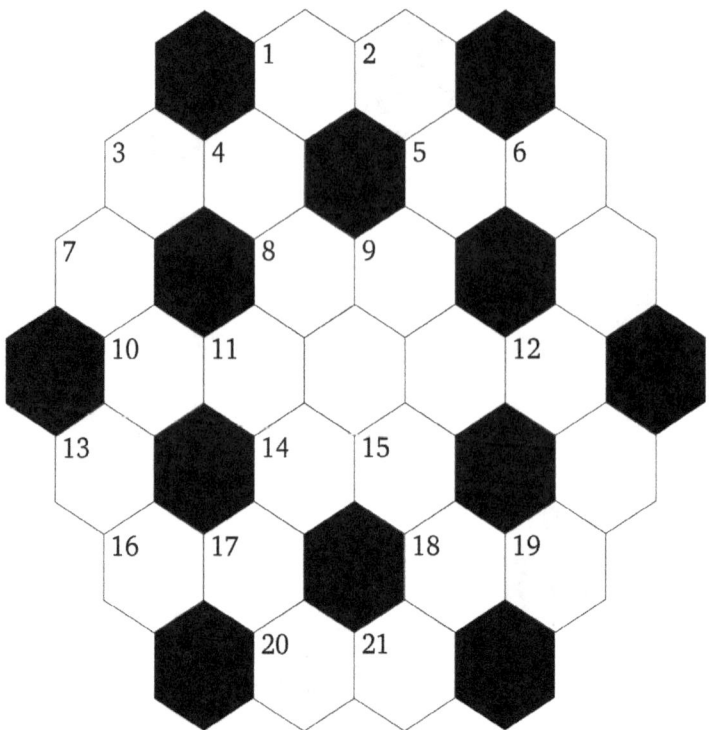

Across

1 Half of 21 up, then subtract 3 across
3 8 across minus 13 down
5 13 up minus 2 down
8 13 up plus 1 across
10 Eighty-seven times a square
14 21 up plus 19 up
16 2 down plus 6 down
18 14 across minus 15 up
20 6 down plus 15 up

Up

4 20 across minus 21 up
7 16 across minus 21 up
11 Mean of 7 up and 11 down
12 17 down minus 5 across
13 16 across minus 19 up
15 14 across minus 7 up
17 Its digits total 19 up
19 13 up minus 17 down
21 16 across minus 18 across

Down

2 14 across minus 4 up
4 Twice a prime number
6 13 up minus 12 up
7 13 down minus 12 up
9 Twice the result of 6 down minus 11 down
11 13 up minus 11 up
12 8 across minus 2 down
13 8 across minus 9 down
17 16 across minus half of 13 up

Puzzle 31

Difficulty: ★★☆☆☆

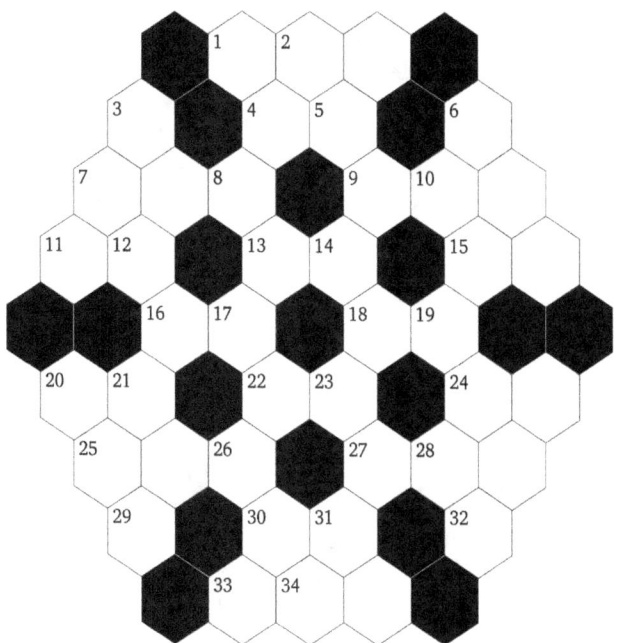

Across

1 3 down plus half of 11 up
4 Mean of 30 across and 31 down
7 Mean of 19 down and 33 across
9 7 across minus 10 down
11 23 up plus 30 across
13 27 across minus 28 up
15 17 down minus 16 across
16 A square
18 11 across minus 33 up
20 4 across minus half of 31 down
22 17 down minus 23 up
24 11 across minus 28 up
25 6 down minus 33 up
27 Mean of 11 across and 20 down
30 11 across minus 33 up
33 8 up plus 27 across

Up

5 23 up minus 28 down
8 Mean of 1 across and 31 down
10 7 across minus 9 across
11 19 down minus 31 down
12 13 across minus 31 down
14 Mean of 17 up and 12 up
17 1 down plus 14 down
19 Sixteen times 14 down
23 11 across minus 18 across
25 6 down plus half of 5 up
26 Mean of 10 up and 18 across
28 27 across minus 29 up
29 Five times 28 down
32 Twice the result of 2 down plus 30 across
33 Mean of 15 across and 20 across
34 11 across times 12 up

Down

1 Mean of 30 across and 10 up
2 1 across minus 8 down
3 1 down plus 28 up
6 23 up plus 25 across
7 24 across plus 9 across
8 5 up plus 28 up
10 Mean of 15 across and 3 down
14 30 across plus 28 up
17 Twice the result of 13 across minus 15 across
19 32 up minus 24 across
20 10 up plus 21 down
21 Five times 20 across
23 31 down minus 14 down
26 33 across plus 22 across
28 21 down divided by five
31 17 up minus 28 up

Puzzle 32

Difficulty: ★★☆☆☆

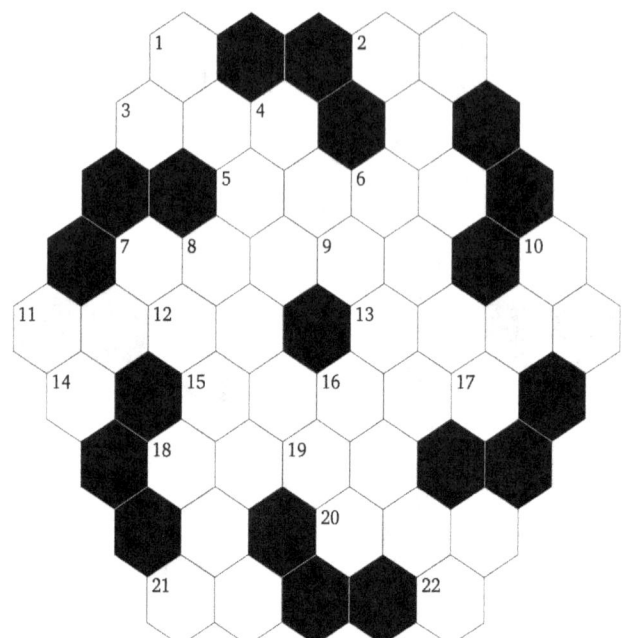

Across

2 Consecutive digits in descending order
3 A prime number
5 A prime number
7 Six times a prime number
11 8 down divided by 3 up
13 A square
15 A prime number
18 Twice a prime number
20 A prime number
21 Seven times a square

Up

3 18 down minus 2 down
9 2 down plus 17 up
12 A prime number
14 A square
17 All digits are the same
18 Twice the result of 4 down minus 17 up
19 Twenty-nine times a square
20 Three hundred ninety-nine less than 12 up
21 Ninety-five times a prime number
22 Half of 1 down, then subtract 16 down

Down

1 Twice a prime number
2 9 up divided by 2 across
4 Three times a prime number
6 Mean of 16 down and 18 across
7 Twice the result of 21 up minus 18 across
8 11 across times 11 down
10 Six times a square
11 Same as 3 up
16 Twelve times a prime number
18 6 down minus 13 across

Puzzle 33

Difficulty: ★★☆☆☆

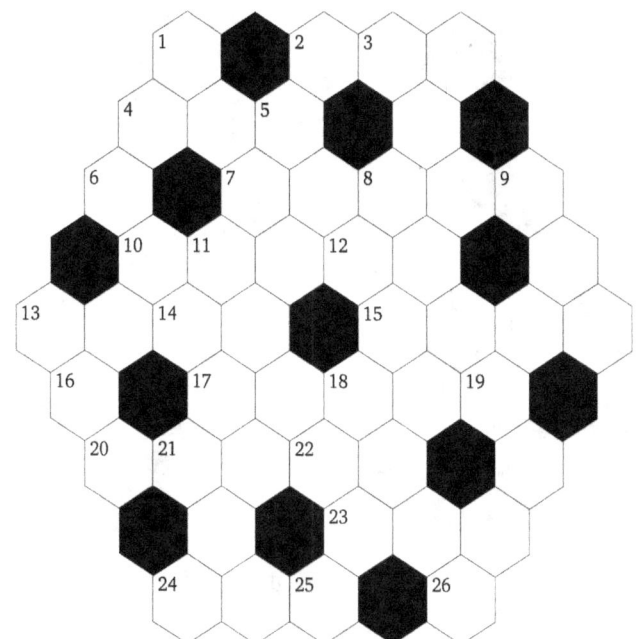

Across

2 Ninety-two more than 3 down

4 Three times a prime number

7 Ten thousand nine hundred sixty less than 8 down

10 Twenty times a prime number

13 26 up plus half of 21 down

15 Three hundred eight more than 1 down

17 Twice a prime number

20 13 down squared

23 Twenty-one times a square

24 A square

Up

6 Mean of 24 across and 19 up

12 Six times a prime number

14 Twenty-seven times a prime number

16 Two hundred nineteen less than 13 across

19 A prime number

21 Thirteen times a prime number

22 Eleven thousand two hundred ninety-six more than 7 across

24 Second, and third digits are the same

25 Mean of 17 across and 15 across

26 Mean of 3 down and 16 up

Down

1 Mean of 18 down and 2 across

3 A square

5 10 across minus 7 across

6 Eighteen thousand four hundred thirty more than 7 across

8 Seven thousand four hundred seventy less than 6 down

9 Twenty-two times a prime number

11 Five times a prime number

13 Three times a prime number

18 Eight times a prime number

21 Twice a prime number

Puzzle 34

Difficulty: ★★☆☆☆

Across

1 2 across reversed
2 7 up minus 19 up
4 Eighty-one times a prime number
10 Thirteen thousand five hundred eighty-two less than 8 down
13 5 up minus 19 up
14 Four times a prime number
15 13 up plus 25 up
16 Twice a prime number
21 8 up minus 22 down
27 Mean of 15 across and 3 down
28 10 across divided by 29 up

Up

5 Mean of 27 across and 3 down
7 Mean of 5 up and 13 across
8 A prime number
13 Mean of 1 across and 20 down
18 25 up plus 15 across
19 Half of 17 up, then subtract 16 across
23 Eighty-nine times a prime number
25 Half of 24 down
26 2 down reversed
29 Sixty less than 12 down

Down

2 Three-fourths of 3 down
3 Mean of 18 up and 25 up
6 5 up minus 26 up
7 Seven times a prime number
8 Twenty-six times a prime number
9 2 across plus half of 8 down
11 15 across plus 18 up
12 A square
20 27 across minus 26 up
22 13 up minus 28 across
24 Mean of 6 down and 3 down

Puzzle 35

Difficulty: ★★☆☆☆

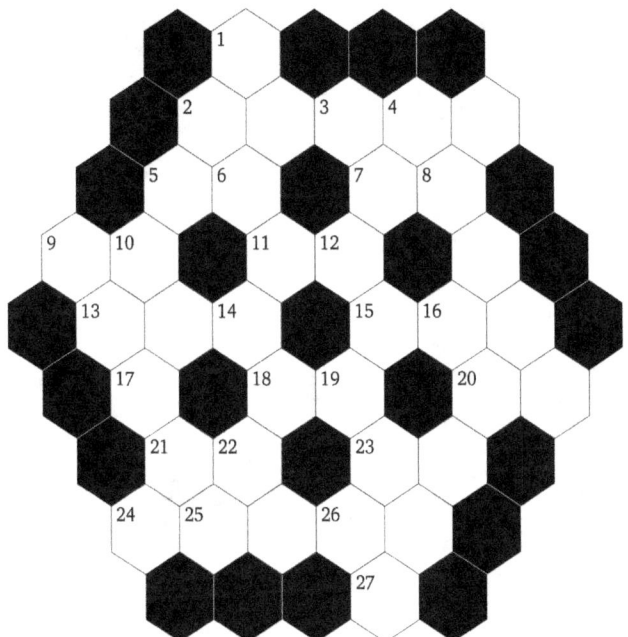

Across

2 Six thousand eight hundred twenty-nine less than 4 down
5 22 down plus 18 across
7 1 down minus 11 across
9 Same as 14 down
11 Mean of 17 up and 8 up
13 26 up plus 5 across
15 19 up plus 1 down
18 11 across plus 12 down
20 Half of 1 down, then subtract 16 up
21 9 across plus 7 across
23 16 up plus 26 up
24 Twice the result of 9 down plus 12 up

Up

6 Twice the result of 13 across minus 18 across
8 1 down minus 17 up
12 15 across plus 25 up
13 Thirty-one times a square
14 5 across minus 8 up
16 26 down minus 3 down
17 23 across minus 26 down
19 6 up minus 9 across
24 23 across plus 11 across
25 Mean of 12 down and 2 down
26 21 across minus 8 up
27 Three times a prime number

Down

1 Mean of 5 across and 9 across
2 26 up plus 19 down
3 Half of 23 across, then subtract 22 down
4 Eighteen thousand four hundred forty-three less than 24 across
9 First two digits are the same as 12 down
10 15 across minus 9 across
12 Mean of 1 down and 22 down
14 15 across minus 10 down
16 23 across minus 14 up
19 13 up divided by 26 down
22 Mean of 3 down and 20 across
26 14 down minus 20 across

Puzzle 36

Difficulty: ★★☆☆☆

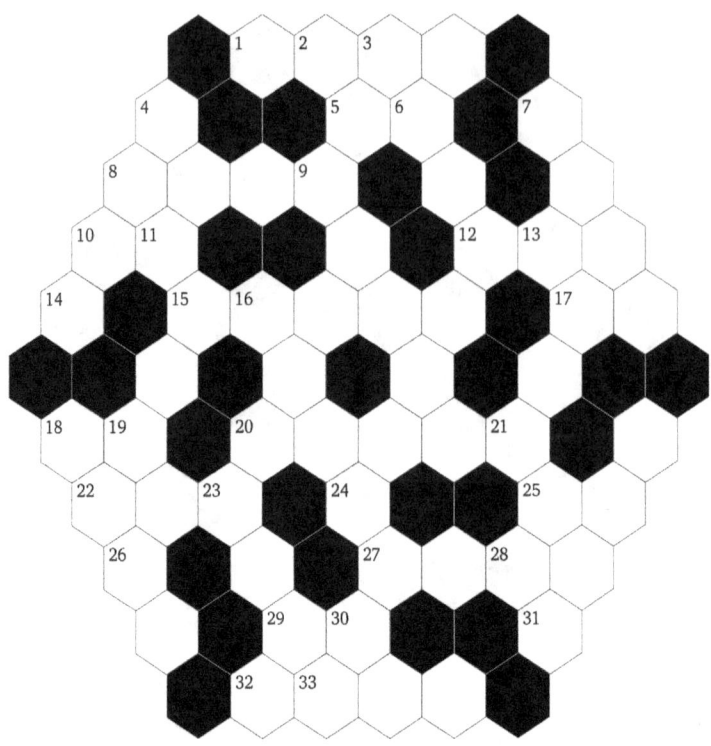

Across

1 Rearranged digits of 23 down
5 32 up plus 10 across
8 Four times a prime number
10 3 down divided by 32 up
12 18 down divided by 17 across
15 A prime number
17 A square
18 10 across minus 29 across
20 A prime number
22 Five times a prime number
25 32 up plus 26 up
27 33 up plus half of 2 down
29 A square
32 Half of 1 across, then subtract 11 up

Up

6 A square
9 A prime number
11 18 across minus 26 up
13 27 across minus 21 down
14 Eighty-four times a prime number
21 Last two digits are the same as last two digits of 27 across
22 A prime number
23 Three hundred thirty-nine less than 20 across
24 Mean of 9 down and 22 across
26 5 across minus 30 down
28 Two-thirds of 32 up
31 14 up minus 23 down
32 Mean of 29 across and 11 up
33 A square

Down

2 21 up divided by ninety-five
3 Ninety-three times 28 up
4 7 down minus 21 down
7 31 up minus 28 down
8 Three times a prime number
9 Three times a prime number
13 28 down plus 19 down
16 A prime number
18 Seventy-two times 19 down
19 Three times 26 up
21 33 up minus 32 up
23 Three times a prime number
28 8 across minus 22 up
30 Mean of 19 down and 4 down

Puzzle 37

Difficulty: ★★☆☆☆

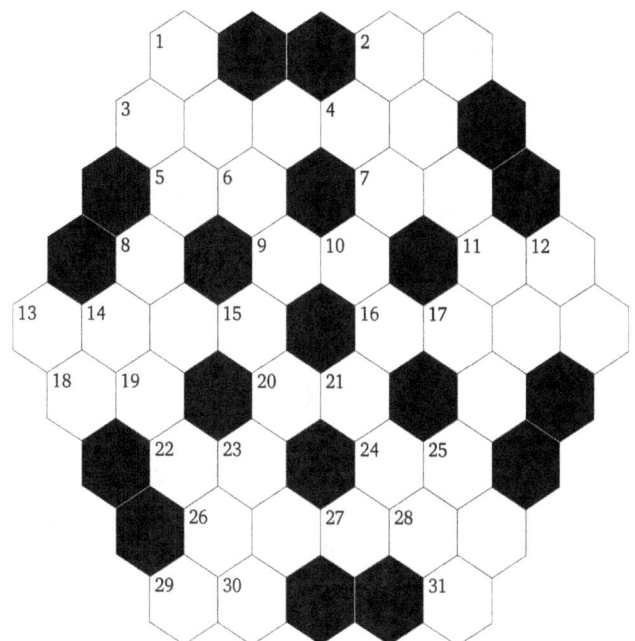

Across

2 Twice the result of 25 down minus 11 across
3 Seven times a prime number
5 Mean of 24 across and 19 up
7 3 down plus 2 across
9 4 up plus 10 down
11 Mean of 5 across and 2 across
13 Forty-eight times a prime number
16 Half of 13 across, then subtract 27 up
18 31 up plus 19 up
20 7 across minus 30 up
22 17 up minus 19 up
24 9 across minus 3 up
26 Nineteen times a prime number
29 Mean of 20 across and 4 down

Up

3 Mean of 18 across and 29 across
4 19 up plus 22 across
6 19 up plus 11 across
10 Thirty times a prime number
15 A square
17 1 down divided by 13 down
18 Twice a prime number
19 3 up reversed
21 Mean of 27 up and 15 up
27 23 down minus 17 up
28 Sixteen thousand four hundred thirty-nine less than 14 down
29 Fifteen times a prime number
30 15 down minus 12 down
31 6 up minus 22 across

Down

1 A square
2 A prime number
3 Same as 13 down
4 7 across minus half of 10 down
8 A square
10 5 across plus 30 up
12 A cube
13 Mean of 21 up and 5 across
14 Nine times a prime number
15 Mean of 15 up and 3 up
17 18 across plus 31 up
21 29 up minus half of 17 down
23 5 across plus 8 down
25 Mean of 4 up and 29 across

Puzzle 38

Difficulty: ★★☆☆☆

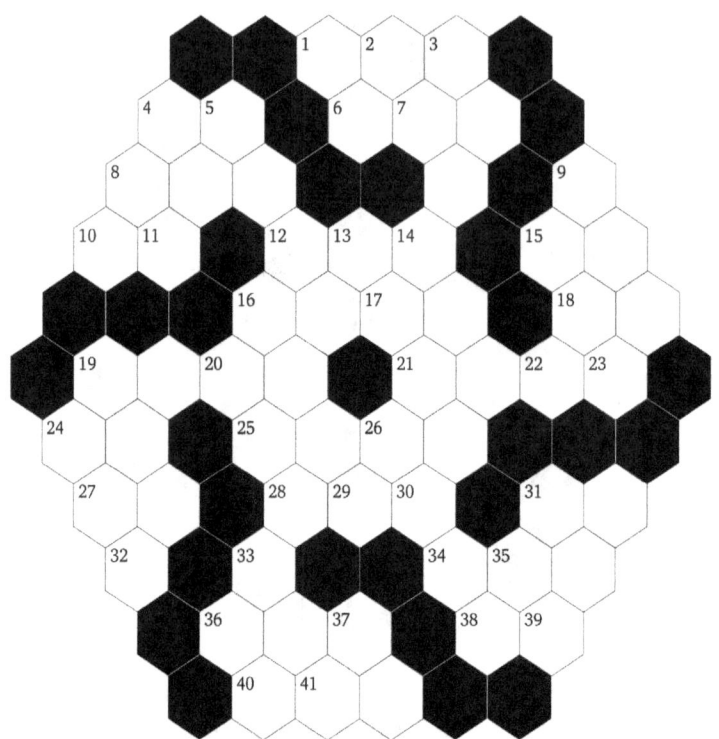

Across

1 39 up minus 31 down
4 20 down minus 38 across
6 27 up plus 33 down
8 A square
10 2 down minus half of 23 up
12 34 across minus 36 down
15 32 up plus 24 across
16 Thirty-six times a prime number
18 Mean of 10 across and 15 across
19 Twice a prime number
21 A cube
24 39 up divided by nine
25 Forty-nine less than 25 up
27 24 across minus 36 down
28 Forty-four less than 19 across

Up

6 Mean of 32 up and 2 down
7 37 down minus 6 up
10 15 down minus 20 up
11 Eighteen times a prime number
15 20 down minus 24 across
17 Mean of 16 down and 36 across
20 27 up minus 14 down
22 35 down plus 12 across
23 20 up minus 10 across
24 Mean of 24 across and 37 down
25 Four times a prime number
27 Twice the result of 30 up minus 14 down
29 Four hundred nine less than 25 across
30 Mean of 37 down and 39 up

Down

1 Mean of 8 down and 38 across
2 31 down plus 3 down
3 9 down divided by six
4 Half of 11 up, then subtract 20 down
5 21 across minus half of 39 up
8 Mean of 31 down and 27 across
9 24 up plus 2 down
13 Twenty-nine times 35 down
14 33 down plus 32 up
15 Five times a prime number
16 Twenty-three times a square
19 4 across plus 15 across
20 6 across minus 30 up
24 22 up plus 3 down
26 Three times a prime number

38

Across (continued)

31 Mean of 18 across and 23 up

34 10 up minus 31 down

36 Mean of 24 down and 14 down

38 Mean of 7 up and 19 down

40 Four times 19 down

Up (continued)

32 25 across divided by 1 across

36 Six times a prime number

38 19 across minus half of 4 across

39 Eight times 15 across

40 Mean of 15 up and 8 down

41 A cube

Down (continued)

31 18 across minus 7 up

33 Twice the result of 9 down minus 1 down

35 Mean of 24 up and 40 up

36 Mean of 4 down and 3 down

37 Mean of 15 up and 36 down

Puzzle 39

Difficulty: ★★☆☆☆

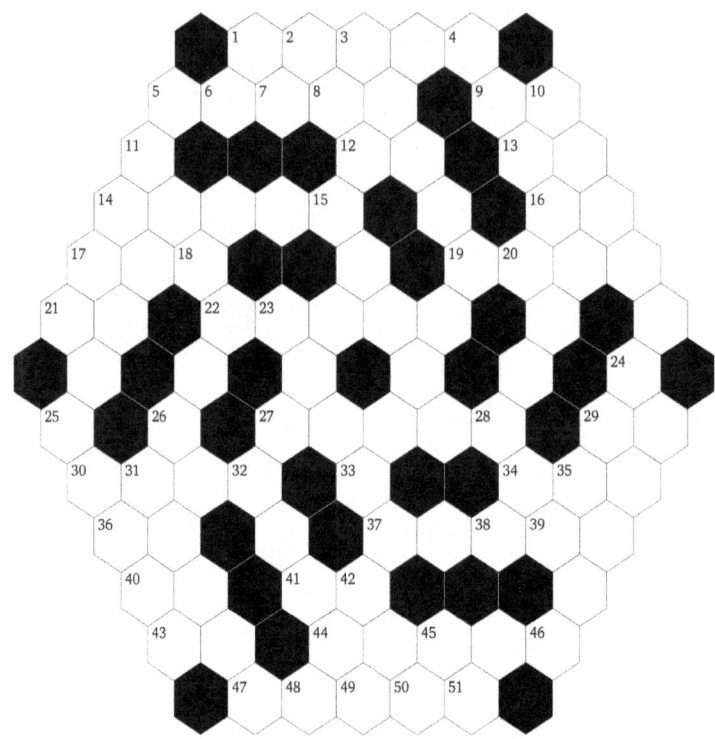

Across

1 Seventy-five times a prime number
5 Last two digits are the same as last two digits of 2 down
9 37 across minus half of 10 down
12 6 up divided by three
13 Half of 50 up, then subtract 24 down
14 27 across minus 40 across
16 50 up minus 40 across
17 49 up plus 9 across
19 Seven times a prime number
21 40 across minus 13 across
22 A prime number
27 A prime number
29 38 up minus 43 across
30 Mean of 19 across and 40 up

Up

6 45 down plus 13 up
7 Mean of 50 up and 29 across
8 42 down minus half of 2 down
13 38 up divided by five
15 16 across times 21 across
18 27 across minus 14 across
20 47 across divided by 38 up
21 A prime number
25 Three thousand six hundred sixty-one less than 32 up
28 Mean of 25 down and 23 down
32 Its digits total 26 down
33 Twenty-five times a prime number
36 A prime number
38 35 down plus 40 across

Down

1 Mean of 18 up and 12 across
2 Twelve times a prime number
3 Two thousand six hundred ten more than 33 up
4 A prime number
10 Four thousand seven hundred sixty less than 5 across
11 34 across times 17 down
14 Mean of 29 across and 36 across
15 8 up plus 22 across
17 Mean of 26 down and 13 up
20 A cube
21 A square
23 Consecutive digits unordered
24 17 down plus 29 down

40

Across (continued)

34 Forty-one times 13 across

36 18 up minus 29 across

37 One thousand two hundred ninety less than 1 across

40 6 up minus half of 7 up

41 20 down minus 13 across

43 Mean of 50 up and 6 up

44 Twenty thousand seven hundred seventy less than 5 across

47 32 down plus 41 across

Up (continued)

39 Three thousand forty-one more than 27 across

40 Five times a square

43 24 down plus 36 across

46 Twice the result of 15 down plus 28 down

48 Sixteen times a prime number

49 8 up minus 35 down

50 18 up plus 16 across

51 43 across reversed

Down (continued)

25 Eight times a prime number

26 40 across divided by four

28 Thirty-two times 7 up

29 12 across minus 26 down

31 Last two digits are the same as last two digits of 30 across

32 47 across minus half of 12 across

35 45 down minus 14 down

42 Twice a prime number

45 36 across plus 26 down

Puzzle 40

Difficulty: ★★★☆☆

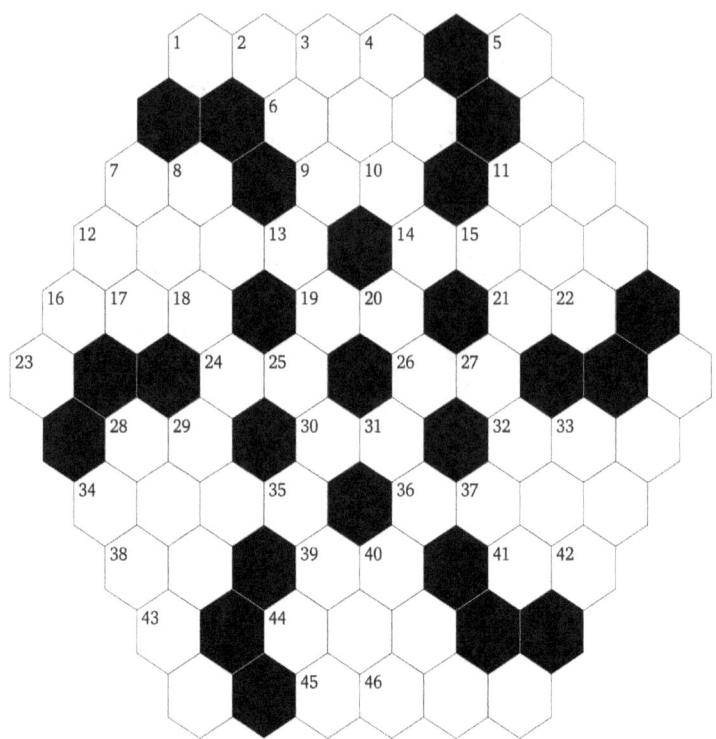

Across

1 Mean of 38 up and 13 up
6 Same as 11 down
7 Twice the result of 40 down minus 6 up
9 Same as 41 across
11 Mean of 25 up and 15 down
12 2 down plus half of 27 up
14 Mean of 42 up and 34 up
16 Fifty-one times 26 across
19 A prime number
21 A square
24 30 across plus half of 12 down
26 35 down divided by 11 down
28 24 across minus 12 down
30 Mean of 15 down and 38 across

Up

6 30 across plus 18 up
10 Same as 25 up
13 14 across minus 27 down
15 Mean of 27 up and 40 down
17 19 across plus 16 across
18 Half of 34 down, then subtract 5 down
20 Mean of 39 across and 19 across
22 20 up plus 18 up
23 3 down minus half of 4 down
25 Mean of 19 across and 9 across
27 Twenty-two times 24 across
31 8 down plus 20 down
34 6 up minus 22 up
35 20 up plus 7 across
37 Mean of 30 across and 24 across

Down

2 12 down times 26 across
3 First two digits are the same as 24 across
4 Mean of 25 up and 44 down
5 Twenty-three times 4 down
7 Twenty-nine times a prime number
8 40 down minus 21 across
11 Twenty-six times a prime number
12 A square
13 11 across minus 25 down
15 37 down plus 18 up
20 14 across divided by 44 up
25 40 down minus 38 across
27 Five times a prime number
28 Eight times 31 up

Across (continued)

32 25 up plus 44 up
34 Fifty-seven times 21 across
36 Consecutive digits unordered
38 34 across divided by 37 down
39 21 across minus 19 across
41 Mean of 28 across and 37 up
44 12 across minus 26 across
45 Mean of 36 across and 44 down

Up (continued)

38 Sixty-one times 6 up
41 Last two digits are the same as 44 down
42 A prime number
43 Half of 28 down
44 32 across minus 7 across
45 Twice a prime number
46 29 down minus 25 up

Down (continued)

29 34 up plus 28 across
31 Mean of 32 across and 18 up
33 Mean of 31 up and 39 across
34 Five hundred seventy-two more than 14 across
35 6 across times 26 across
37 9 across plus 26 across
40 18 up plus 44 down
44 8 down plus 7 across

Puzzle 41

Difficulty: ★★★☆☆

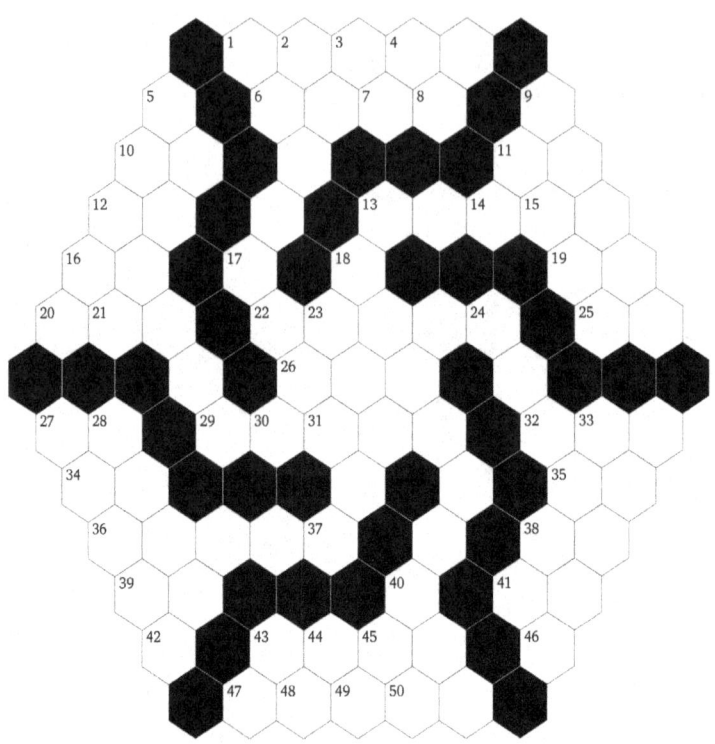

Across

1 A prime number
6 Fifty-one times a square
10 44 down plus 34 across
11 26 across minus 34 across
12 Mean of 2 down and 27 across
13 47 across plus half of 12 down
16 38 across minus 44 down
19 20 across minus 25 up
20 A square
22 Three thousand four hundred forty-six more than 1 across
25 44 down plus 19 across
26 48 up plus 36 up
27 Mean of 34 across and 3 down
29 First two digits are the same as 27 across
32 40 down minus 1 down

Up

6 16 across plus 47 up
7 48 up plus 33 down
8 Same as 16 across
14 A square
15 5 down plus 19 across
17 A prime number
19 Three-fourths of 15 up
20 Mean of 24 down and 15 up
21 Twenty-five times 1 down
25 Twice 27 across
30 A prime number
31 48 up plus 36 up
34 Mean of 15 up and 5 down
36 34 across minus 5 down
37 Six thousand ninety-five more than 47 across
39 Twice the result of 20 across minus 6 up
41 Twenty-four times a prime number

Down

1 6 up plus 39 up
2 34 across plus 4 down
3 23 down minus 25 across
4 Mean of 3 down and 36 up
5 Twice the result of 45 down minus 34 up
9 Seven times a prime number
10 17 up minus 6 across
11 19 up times 48 up
12 First two digits are the same as 16 across
16 A square
17 30 up plus half of 41 up
18 Mean of 1 across and 9 down
23 16 across plus half of 25 up
24 Last two digits are the same as 38 across

Across (continued)

34 25 across minus 4 down
35 Same as 10 across
36 Twice the result of 46 up plus 35 across
38 26 across minus 41 down
39 44 down plus 48 up
41 A square
43 27 across times 47 up
47 Four thousand seven hundred seventy-eight more than 36 across

Up (continued)

42 14 up minus 11 across
46 Three thousand four hundred seventy-three more than 9 down
47 Mean of 11 across and 33 down
48 39 up minus 33 down
49 Ten thousand fourteen more than 12 down
50 Mean of 31 up and 43 down

Down (continued)

27 Two thousand five hundred thirty-six more than 37 up
28 37 up divided by 36 up
33 25 up minus 19 across
38 Six times 41 across
40 42 up plus half of 4 down
41 44 down minus 41 across
43 25 across minus 48 up
44 38 across minus 8 up
45 38 down minus 33 down

Puzzle 42

Difficulty: ★★★☆☆

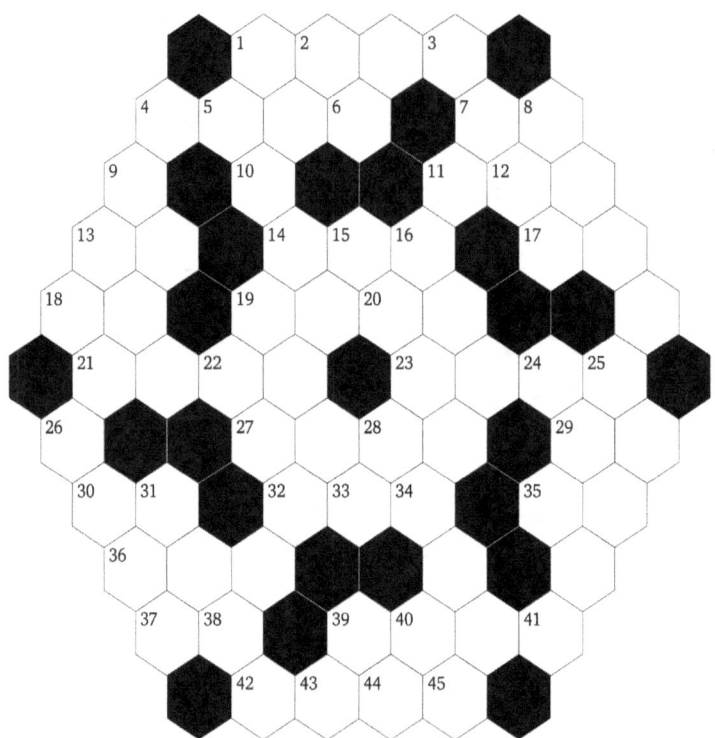

Across

1 Thirty-six times a prime number

4 Twice a prime number

7 23 across divided by 18 across

11 7 across plus 36 across

13 44 up divided by eighteen

14 Nine times 6 up

17 5 up plus 18 down

18 23 across divided by 35 down

19 First two digits are the same as 6 up

21 Mean of 32 across and 19 down

23 7 across times 18 down

27 Fifty-three times 24 down

29 2 down minus 30 across

30 35 down minus half of 36 up

32 27 across minus 8 down

Up

5 Mean of 18 down and 9 down

6 30 across plus 13 across

10 19 across minus half of 3 down

12 25 down plus 39 down

17 A square

18 21 across plus 45 up

20 15 down minus 35 down

22 Twice the result of 14 across plus 35 across

26 Thirty-six times 13 across

27 Two hundred fifty-four more than 28 down

33 Ninety-two times a prime number

34 A square

35 A prime number

36 A square

37 24 down minus 6 up

Down

1 Mean of 24 down and 5 up

2 Mean of 6 up and 25 down

3 Fifty-four times a prime number

5 33 up minus 36 up

8 28 down minus 29 across

9 Mean of 45 up and 1 down

13 Mean of 36 across and 40 down

15 20 up plus 7 across

16 Eleven times 25 down

18 39 down minus 13 across

19 42 across minus 13 across

22 23 across minus half of 34 up

24 16 down minus 34 up

25 28 down minus 8 down

46

Across (continued)

35 24 down minus 17 across

36 A square

37 Mean of 34 up and 7 across

39 Nine times a prime number

42 19 down plus 2 down

Up (continued)

38 Thirteen times a prime number

41 A square

43 17 up minus 35 down

44 Half of 26 up

45 36 up plus 43 up

Down (continued)

26 27 across minus 39 across

28 Twenty-seven times a prime number

31 5 up times 37 across

35 Mean of 13 across and 5 up

39 2 down plus 18 down

40 44 up minus 10 up

Puzzle 43

Difficulty: ★★★☆☆

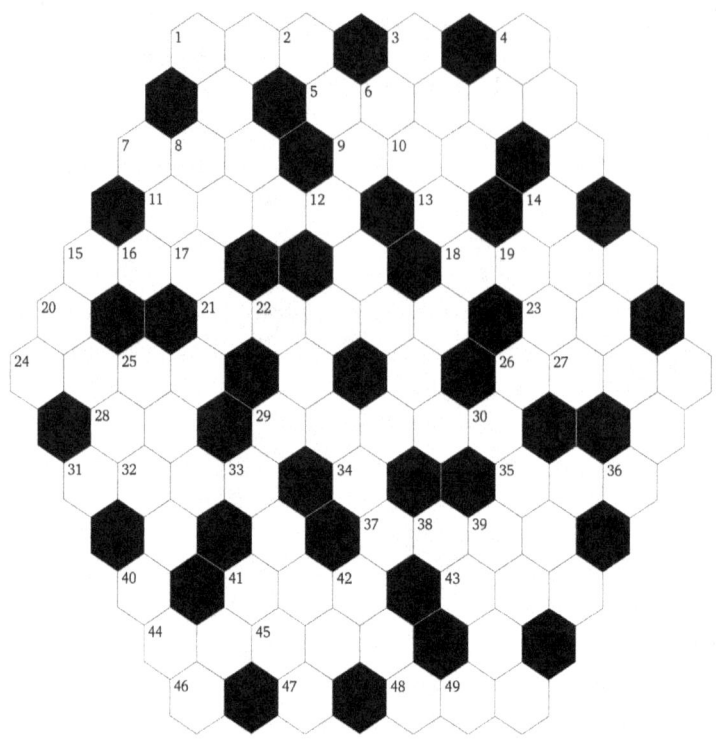

Across

1 43 across plus 39 down
5 Twenty-six times a prime number
7 13 up minus half of 1 down
9 8 down plus 41 across
11 A prime number
15 Four times 23 across
18 Three times a prime number
21 Sixty-four times a prime number
23 Half of 39 down
24 One thousand four hundred twenty-five more than 6 down
26 Twice a prime number
28 30 up divided by ninety-six
29 A prime number
31 First two digits are the same as first two digits of 3 down

Up

10 A square
12 33 down plus 43 up
13 A prime number
16 A prime number
17 Three times a prime number
19 24 up minus 23 across
24 One hundred eighty-three more than 2 down
27 7 down minus half of 35 across
30 41 down plus 42 down
31 Same as 1 across
32 47 up minus 42 down
33 Twice a prime number
34 Twice the result of 22 down plus 19 up
36 Mean of 41 down and 40 down
40 A prime number
43 Twenty-nine times a prime number

Down

1 Forty more than 31 across
2 31 across divided by 28 across
3 15 across plus 41 across
4 Mean of 48 across and 8 down
6 Twice a prime number
7 A prime number
8 45 up minus 28 across
12 A prime number
14 A prime number
19 Seven times a square
20 A prime number
22 Forty-eight times a prime number
25 Mean of 17 up and 9 across
30 Twice a prime number
33 Twice the result of 49 up minus 24 across
38 Twice a prime number
39 Half of 15 across

Across (continued)

35 One hundred five more than 24 up

37 6 down minus half of 36 up

41 Five times 7 across

43 31 up minus 39 down

44 33 up minus 24 up

48 Rearranged digits of 27 up

Up (continued)

45 Four times a prime number

46 4 down plus 36 up

47 A square

49 Thirty-nine times a prime number

Down (continued)

40 46 up minus half of 1 across

41 One hundred forty-nine less than 1 across

42 27 up minus 15 across

Puzzle 44

Difficulty: ★★★☆☆

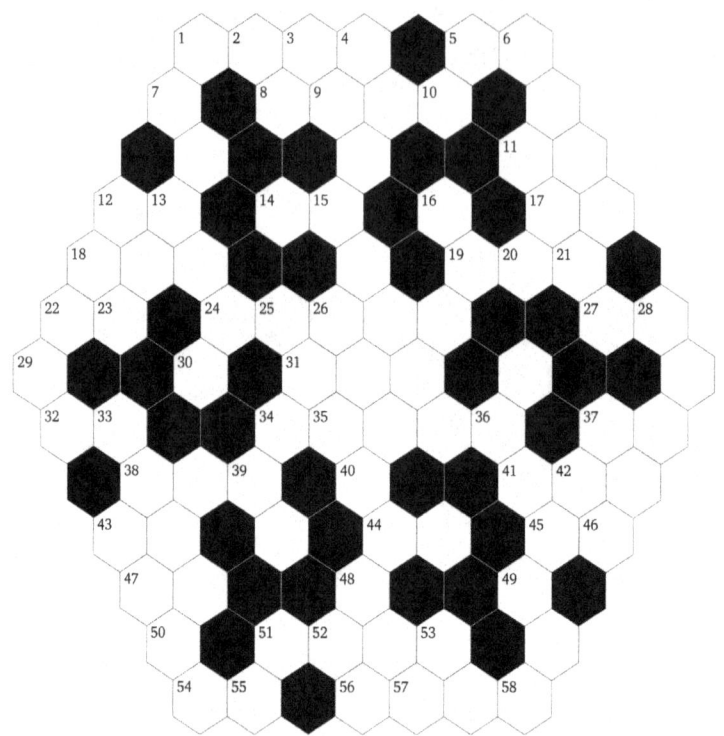

Across

1 6 down plus 38 across
5 55 up minus half of 28 down
8 Thirty-three times a prime number
11 Mean of 2 down and 49 down
12 Mean of 48 down and 52 down
14 10 up plus 12 across
17 A square
18 Rearranged digits of 13 down
19 35 up minus 44 across
22 42 down plus 7 up
24 25 down plus 27 across
27 4 down minus 54 across
31 Seventy-six times 16 down
32 Mean of 16 down and 15 up
34 Seventy-nine times a prime number

Up

7 32 across minus 21 up
8 19 across minus half of 26 down
9 16 down plus half of 37 down
10 52 up minus 15 up
11 18 down minus 54 across
15 50 up plus 17 across
20 Six times 18 down
21 43 across plus 43 up
23 15 down minus half of 50 up
29 47 up times 7 up
30 54 across plus 53 down
35 5 across times 56 up
36 14 across plus 43 across
39 Twenty times a prime number
40 Last two digits are the same as 16 down
43 42 down minus 29 down

Down

2 A square
3 Fifty times 7 down
4 47 across plus 54 across
6 Mean of 5 across and 43 down
7 31 across divided by forty-eight
11 33 down plus 56 up
12 4 down minus 43 up
13 3 down minus 14 across
15 Twice a prime number
16 12 across minus 21 up
18 Mean of 48 down and 11 up
25 24 across minus 50 up
26 45 across plus 38 across
28 10 up plus 8 up
29 53 down minus half of 17 across
33 Fifty-six times a prime number
36 21 up plus 13 down
37 Same as 37 across

Across (continued)

37 Twice the result of 36 down minus 13 down

38 26 down minus 12 across

41 Mean of 35 up and 19 across

43 42 down minus 29 down

44 Two-thirds of 49 down

45 Mean of 7 up and 48 down

47 24 across minus 25 down

51 Sixty-eight times 5 across

54 31 across divided by 32 across

56 Ninety-six times 54 across

Up (continued)

46 Twice a prime number

47 29 up divided by 52 down

49 Twenty-one times a prime number

50 18 down minus 43 up

52 43 up plus 37 down

55 Mean of 47 across and 12 across

56 53 down minus 11 up

57 Mean of 48 down and 52 down

58 Twice the result of 48 down minus 11 up

Down (continued)

39 Sum of digits in 39 up

42 Mean of 36 up and 4 down

43 Three times a prime number

48 7 up plus 36 up

49 Mean of 47 across and 15 up

52 30 up minus 42 down

53 22 across minus 54 across

Puzzle 45

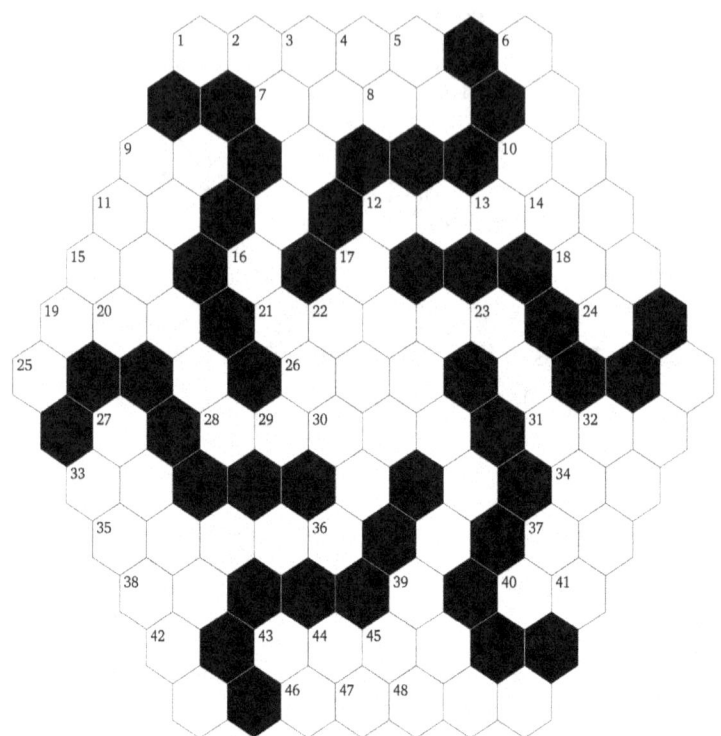

Across

1 35 across minus 16 down
7 6 down minus half of 31 across
9 39 down minus 46 up
10 48 up divided by six
11 38 up minus 10 across
12 Twice a prime number
15 Same as 44 down
18 Mean of 15 across and 32 down
19 Eleven times 18 across
21 Half of 23 down, then subtract 22 down
26 Five times 39 down
28 Eight times a prime number
31 Six times 7 up
33 A square
34 7 up minus 33 up
35 Six times a prime number
37 Mean of 3 down and 18 across

Up

7 32 down plus 33 up
8 14 up plus 34 across
13 Twenty-one times 10 across
14 35 up minus 24 up
16 Two thousand twenty-three more than 28 across
18 40 across plus 10 across
20 Three times a prime number
24 18 across minus 10 across
25 15 down plus half of 28 across
29 Twice the result of 33 down plus 33 across
30 Four times 44 down
33 Twice the result of 17 down minus 36 up
35 Mean of 45 down and 7 up

Down

2 3 down plus 26 across
3 43 down plus 46 up
4 8 up plus 10 across
5 9 across minus 10 across
6 Last two digits are the same as last two digits of 10 down
9 32 down plus 43 down
10 Twenty-three times 24 up
11 12 across plus 16 up
15 Mean of 45 down and 18 up
16 Eight times a prime number
17 Thirteen thousand nine hundred thirty-nine less than 11 down
22 First two digits are the same as 7 up
23 Thirty-four times a prime number
27 A prime number

Across (continued)

38 43 down plus 34 across
40 Same as 45 down
43 Fifty times a prime number
46 38 across plus half of 47 up

Up (continued)

36 Twenty-seven times a prime number
38 46 up plus 5 down
40 First two digits are the same as 24 up
41 Its digits total 43 down
42 33 up plus 31 across
46 A square
47 Four times a prime number
48 Mean of 43 down and 11 across

Down (continued)

32 9 across minus 33 across
33 Last two digits are the same as 45 down
37 Twice the result of 20 up minus 10 down
39 11 across plus 14 up
43 38 up minus 34 across
44 Mean of 34 across and 40 across
45 Same as 37 down

Puzzle 46

Difficulty: ★★★☆☆

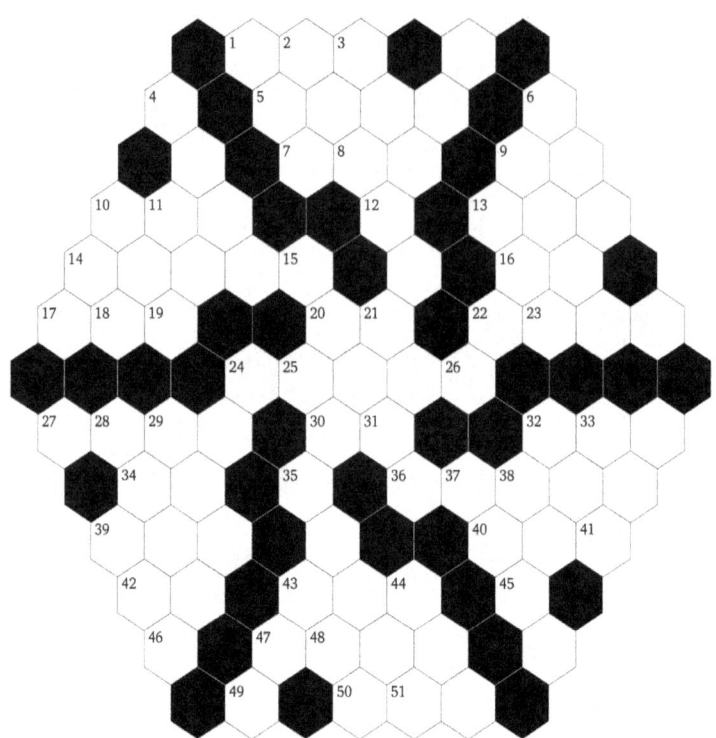

Across

1 40 up minus 10 down
5 Twenty-five times a prime number
7 17 across minus 21 down
9 39 down minus 7 across
10 27 across divided by 25 up
13 Mean of 1 down and 5 up
14 Twenty-seven times a prime number
16 37 down divided by 14 down
17 Same as 39 across
20 Mean of 34 across and 42 across
22 Three times a prime number
24 A prime number
27 Eighty-four times 48 up
30 Two-fifths of 1 down
32 43 down plus 14 down

Up

5 13 up minus 46 up
7 39 down minus 25 up
8 13 up minus 10 down
12 Thirty-four times a prime number
13 38 down plus 34 across
17 Seven times 20 across
18 Two hundred thirty-seven less than 4 down
19 39 across minus 21 down
23 20 across plus 11 down
25 16 across minus 48 up
26 Ten times a prime number
31 A square
35 Mean of 2 down and 22 across
39 Mean of 39 down and 39 across
40 50 up plus 5 up

Down

1 37 down divided by 21 down
2 Five thousand nine hundred sixty-two less than 14 across
3 23 up plus 1 down
4 44 down times 8 up
6 43 across minus 17 across
9 18 up minus 10 down
10 A cube
11 Mean of 9 across and 30 across
13 42 across plus 51 up
14 10 across minus 16 across
15 7 across times 11 down
21 12 up divided by 13 down
25 27 across divided by 10 across
28 Twelve times a prime number

Across (continued)

34 29 down minus 43 down

36 Six times a prime number

39 43 across minus 6 down

40 Mean of 19 up and 20 across

42 9 across plus 21 down

43 Mean of 41 up and 43 down

47 9 down minus 7 up

50 Mean of 39 down and 13 down

Up (continued)

41 Mean of 29 down and 43 down

42 A prime number

45 A prime number

46 11 down plus 34 across

48 Mean of 25 up and 14 down

49 28 down minus half of 50 across

50 Mean of 13 down and 17 up

51 Three-fifths of 46 up

Down (continued)

29 15 down minus 36 across

32 Twenty-seven times a prime number

33 34 across minus 21 down

35 36 across minus 9 down

37 30 across times 16 across

38 1 down minus 10 down

39 43 down minus 13 up

43 32 across minus 30 across

44 14 down plus 38 down

Puzzle 47

Difficulty: ★★★☆☆

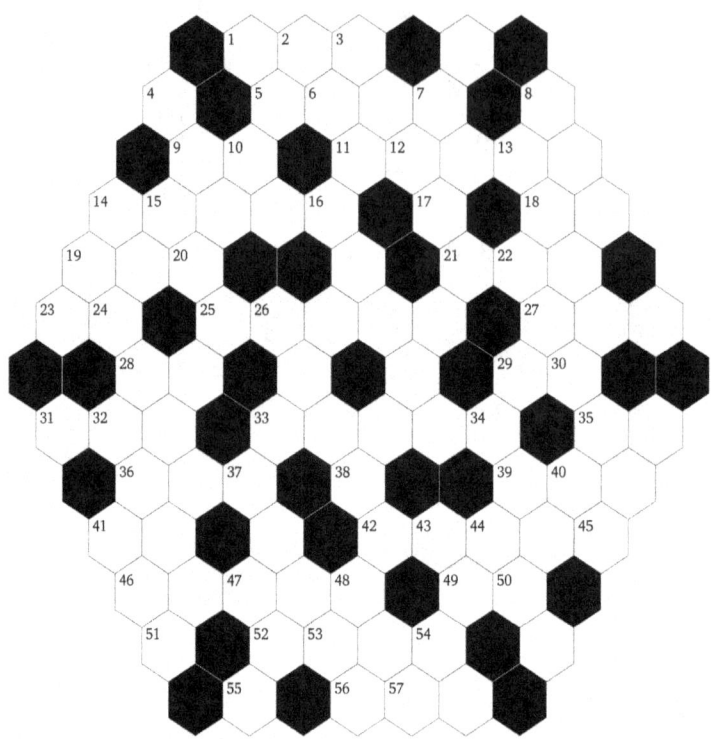

Across

1 Its digits total 1 down
5 Sixty-three times 28 across
9 39 across minus 7 down
11 24 up times 56 up
14 Thirty-one times a prime number
18 Mean of 30 up and 47 down
19 4 down minus 18 across
21 56 across plus 51 up
23 9 across minus 40 down
25 Four times a prime number
27 Sixteen times a prime number
28 1 across minus 8 down
29 46 up minus 55 up
31 56 up plus 22 up
33 Mean of 25 across and 39 across
35 41 across plus 56 up

Up

6 A square
12 Fifty-six times a prime number
13 Mean of 7 down and 9 across
16 Mean of 23 up and 21 across
17 45 up minus 34 down
20 41 down times 10 down
22 21 across plus 29 across
23 16 up plus 39 across
24 Eight times a prime number
30 39 across minus 23 across
32 8 down minus 17 up
34 A prime number
37 Twenty-three times a prime number
38 Twice the result of 26 down minus 31 across

Down

1 14 down minus 47 down
2 55 up plus 27 across
3 Twice the result of 16 down minus 28 across
4 Twelve times 47 down
7 30 up minus 40 down
8 21 across minus 39 across
10 36 across minus 53 up
13 Six times a prime number
14 Mean of 13 up and 10 down
15 23 up minus 41 across
16 Seven thousand six hundred seventy-four less than 33 across
19 Three times a prime number
22 Six hundred ninety less than 38 up
26 Seven times a prime number

Across (continued)

36 Rearranged digits of 15 down

39 45 up minus 48 down

41 14 down plus 56 up

42 Mean of 22 down and 13 down

46 56 across times 32 up

49 23 across minus 40 down

52 Seventy-two times 47 down

56 46 across divided by 54 down

Up (continued)

41 Two thousand six hundred twenty-five more than 37 up

45 7 down plus half of 12 up

46 56 across plus 1 down

47 A prime number

50 32 down plus 2 down

51 41 across minus 23 across

53 27 across minus 29 across

55 Mean of 46 up and 35 across

56 32 up minus 30 up

57 Eight thousand eight hundred eighty-seven less than 41 up

Down (continued)

32 Five times a prime number

34 Three-fourths of 27 across

37 First two digits are the same as first two digits of 41 down

40 A square

41 44 down plus 1 down

43 Mean of 9 across and 7 down

44 36 across plus 43 down

47 A square

48 Mean of 1 down and 44 down

54 56 up plus 30 up

Puzzle 48

Difficulty: ★★★☆☆

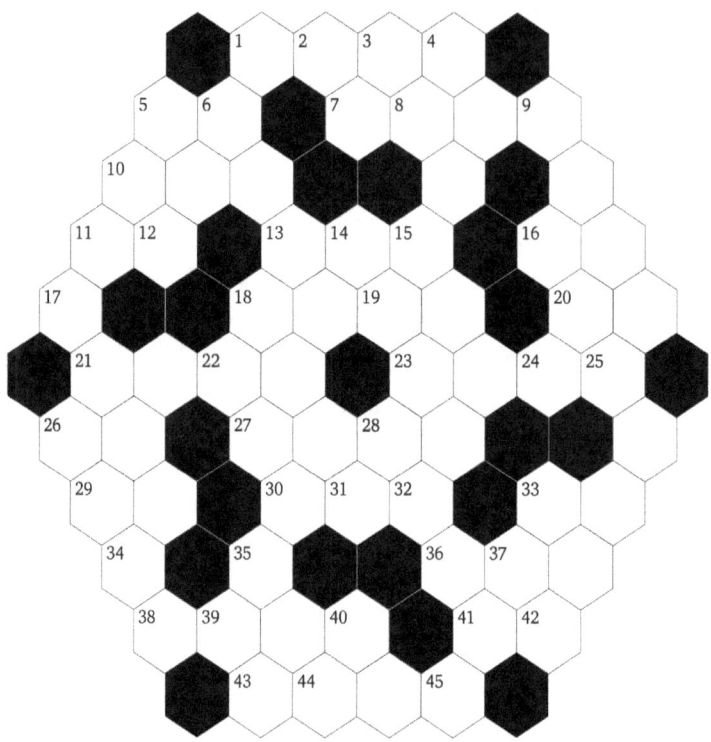

Across

1 Rearranged digits of 6 down
5 30 across divided by 5 down
7 A prime number
10 Three times a prime number
11 26 up plus 40 down
13 Thirty-seven times 5 down
16 11 across plus 5 across
18 Twice the result of 21 across minus 2 down
20 37 down plus 26 across
21 26 up plus 26 down
23 Half of 27 across, then subtract 36 across
26 22 down divided by 5 across
27 Three hundred forty-nine less than 43 across
29 20 across plus 26 across

Up

7 11 across minus 40 down
8 2 down minus 5 down
12 A prime number
16 Same as 25 up
17 Six hundred thirty-five more than 17 down
19 Half of 16 down, then subtract 17 up
22 32 up minus 2 down
24 Three times a prime number
25 34 up plus 44 up
26 43 up minus 10 down
27 45 up plus 15 down
29 6 down minus 22 up
31 Thirty-five times 22 up
32 Three times a prime number
34 22 up divided by 26 across
39 Six hundred thirty-five less than 43 across

Down

2 10 down minus 34 up
3 11 across plus 33 across
4 33 down plus 34 up
5 39 down minus 44 up
6 A prime number
9 Ninety-three times a prime number
10 Mean of 16 up and 20 across
14 One thousand one hundred eighty-one more than 23 across
15 5 across plus 10 down
16 One hundred eleven more than 43 across
17 Three times a prime number
18 Fifty-seven times a square
22 Mean of 24 up and 39 down
26 1 across plus 35 down
28 A cube

Across (continued)

30 Seven times 37 down
33 Mean of 10 down and 8 up
36 Two-fifths of 26 down
38 Four times a prime number
41 Mean of 29 across and 5 across
43 42 up plus 43 up

Up (continued)

42 43 across minus 39 down
43 44 up plus 5 down
44 18 down divided by 8 up
45 One thousand two hundred thirty-six more than 31 up

Down (continued)

33 Two-thirds of 43 up
35 10 down plus 15 down
37 Mean of 2 down and 40 down
39 5 across plus 11 across
40 Mean of 33 down and 33 across

Solutions

Puzzle 1

Puzzle 2

Puzzle 3

Puzzle 4

Puzzle 5

Puzzle 6

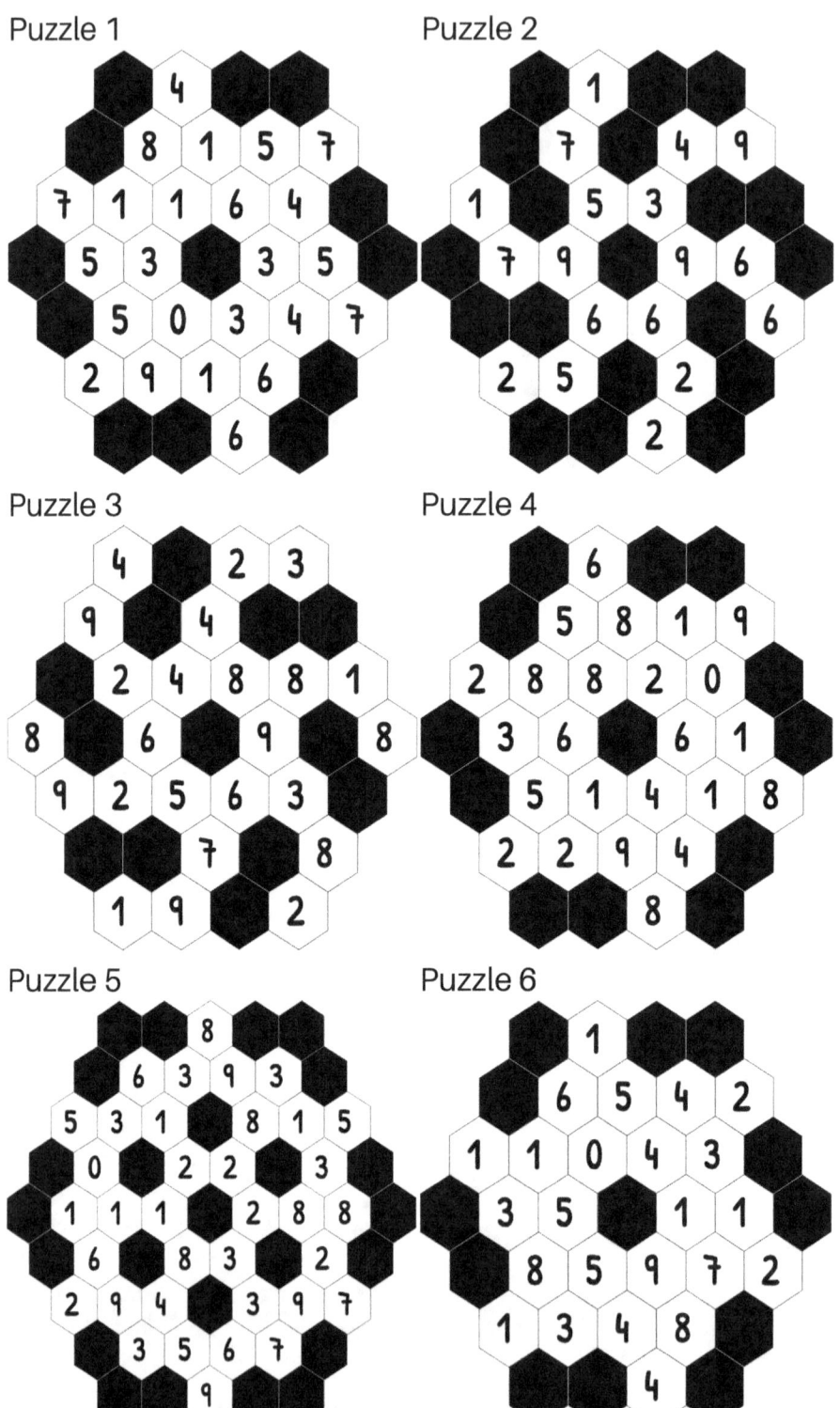

Puzzle 7

Puzzle 8

Puzzle 9

Puzzle 10

Puzzle 11

Puzzle 12

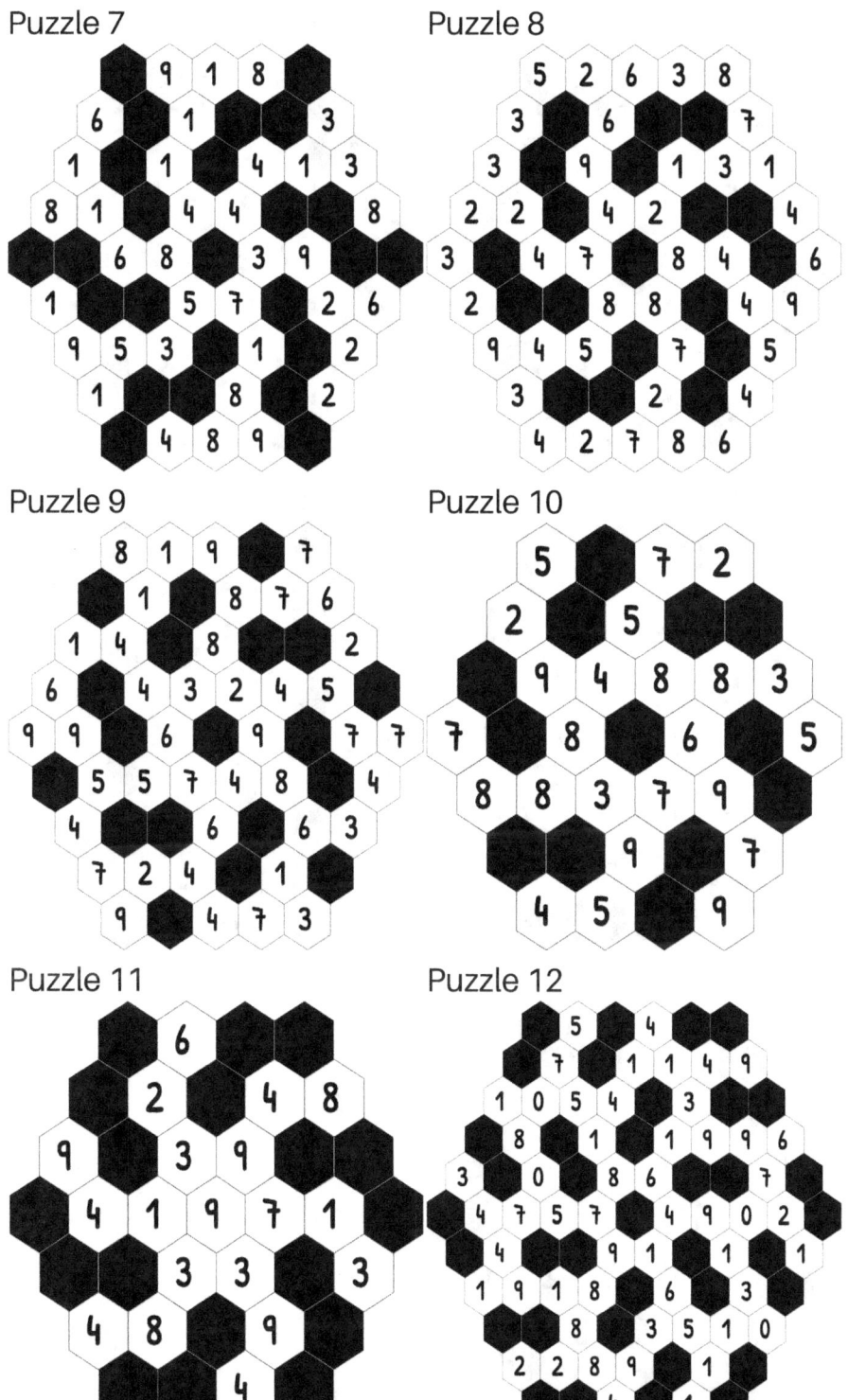

63

Puzzle 13

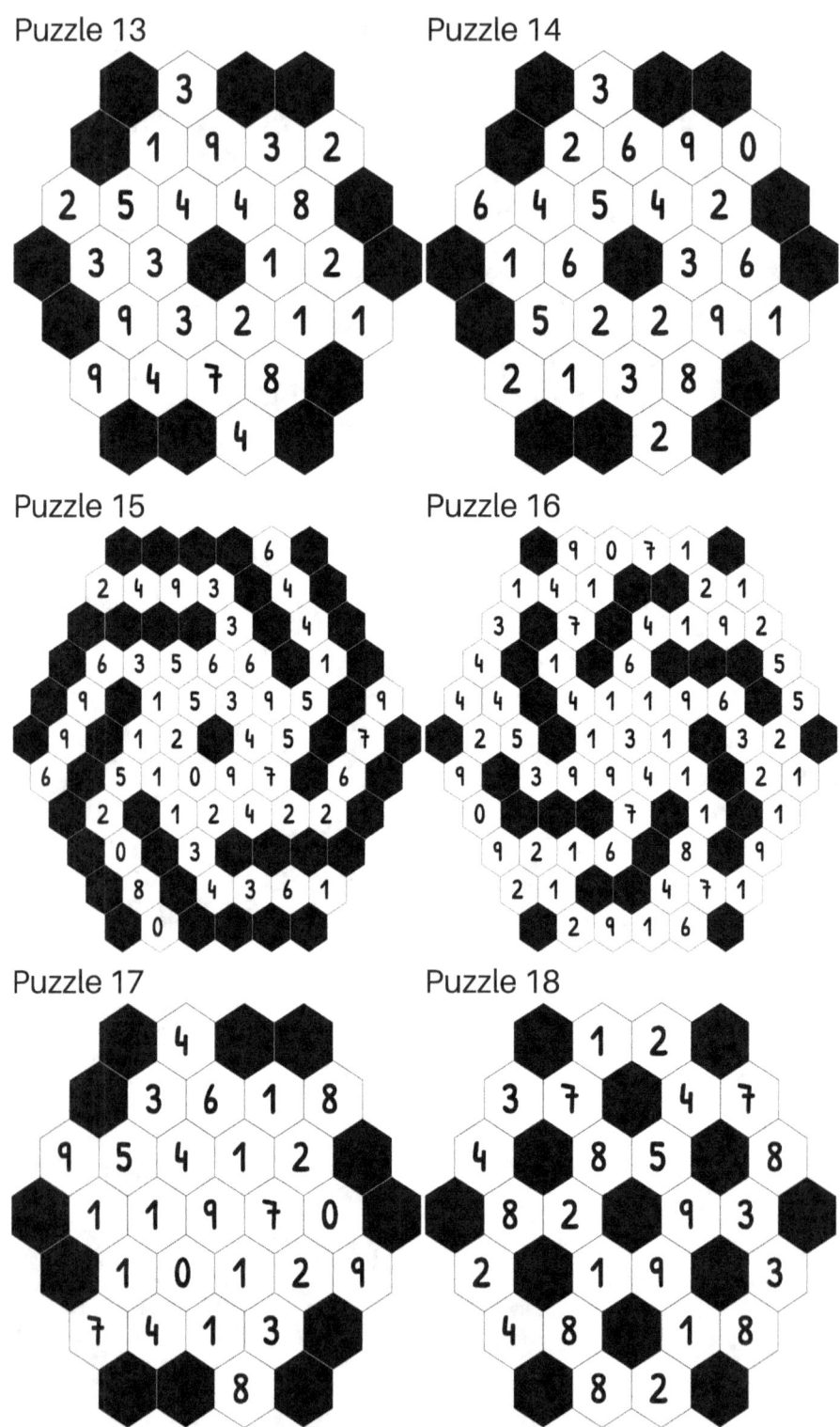

Puzzle 14

Puzzle 15

Puzzle 16

Puzzle 17

Puzzle 18

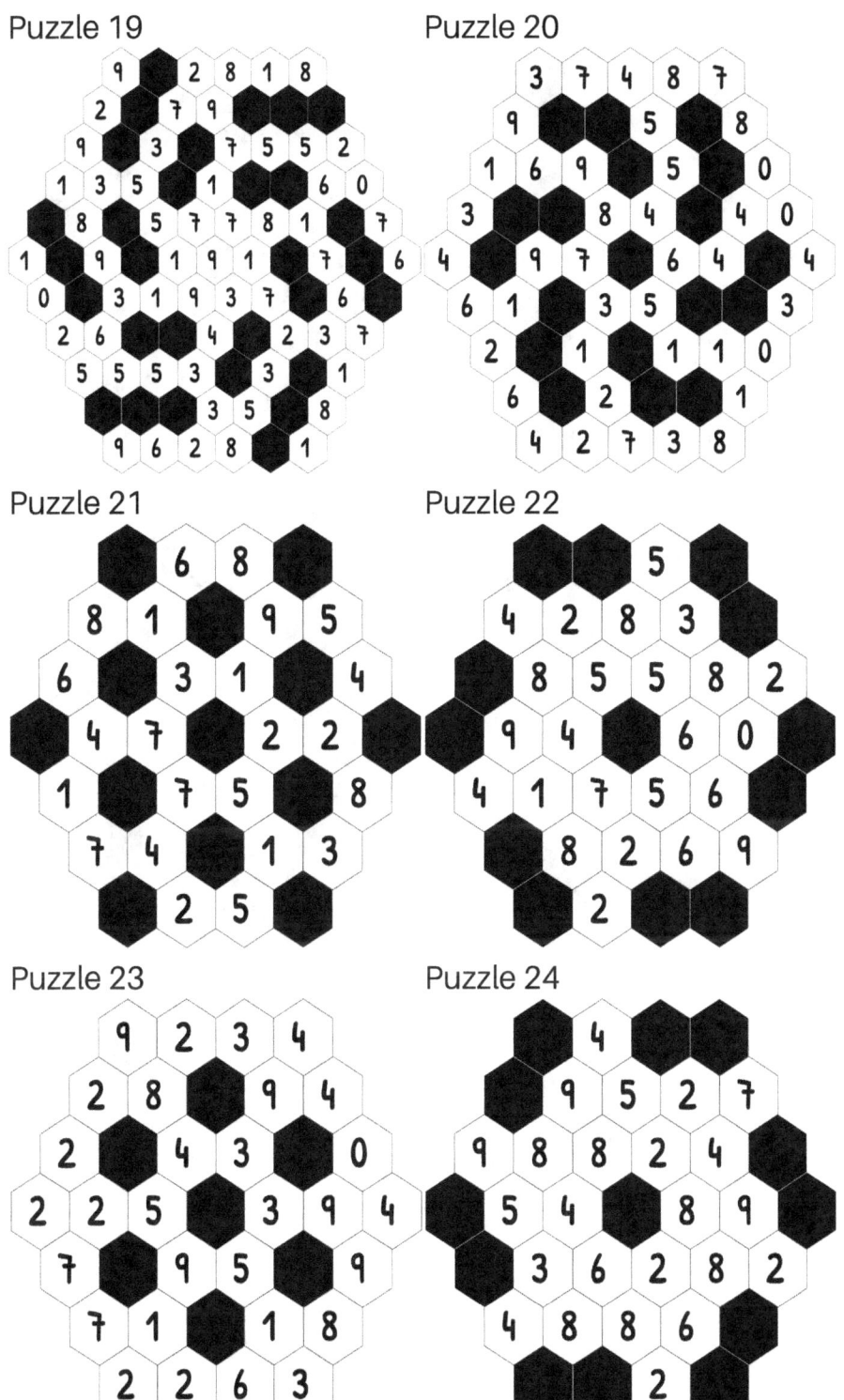

65

Puzzle 25

Puzzle 26

Puzzle 27

Puzzle 28

Puzzle 29

Puzzle 30

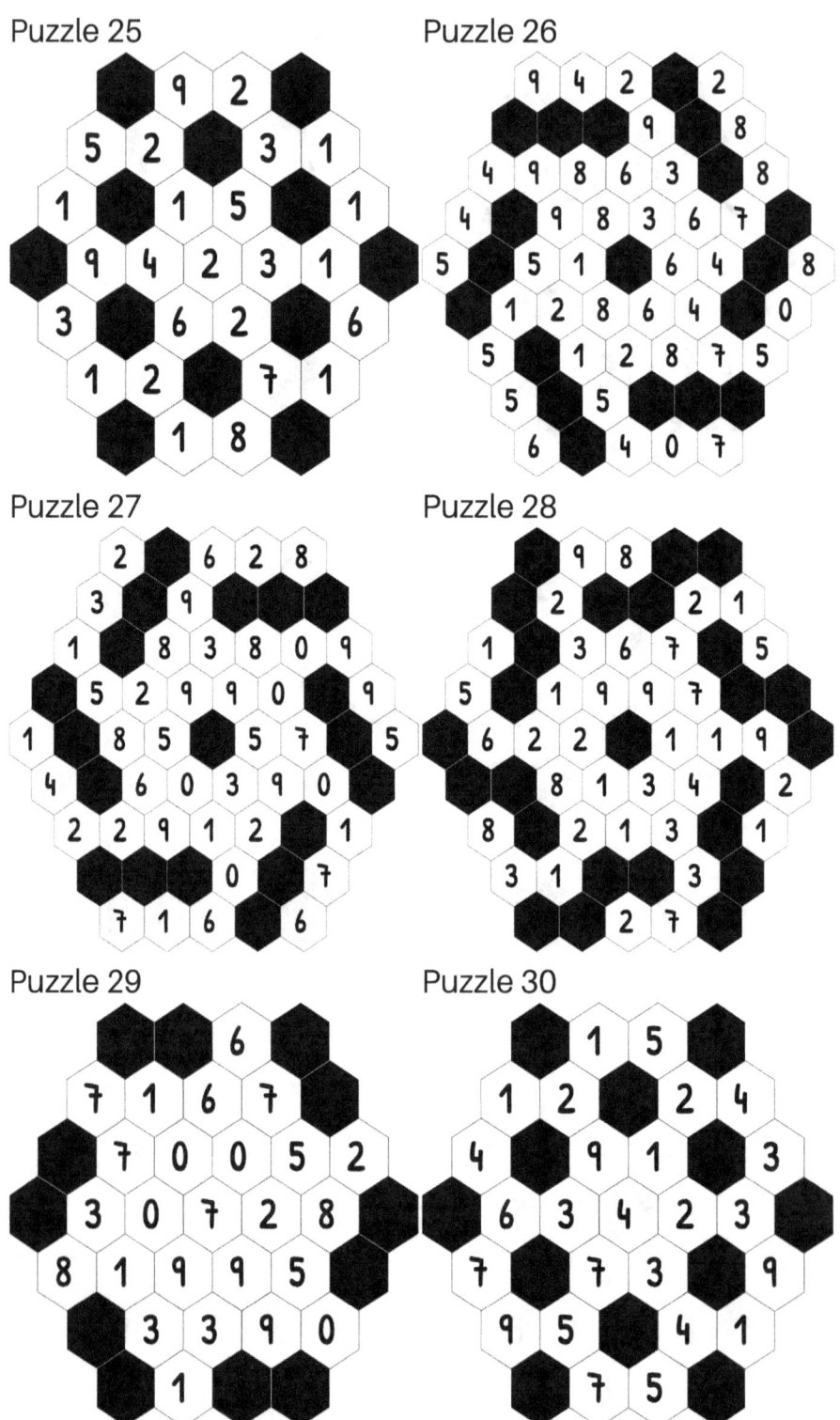

Puzzle 31

Puzzle 32

Puzzle 33

Puzzle 34

Puzzle 35

Puzzle 36

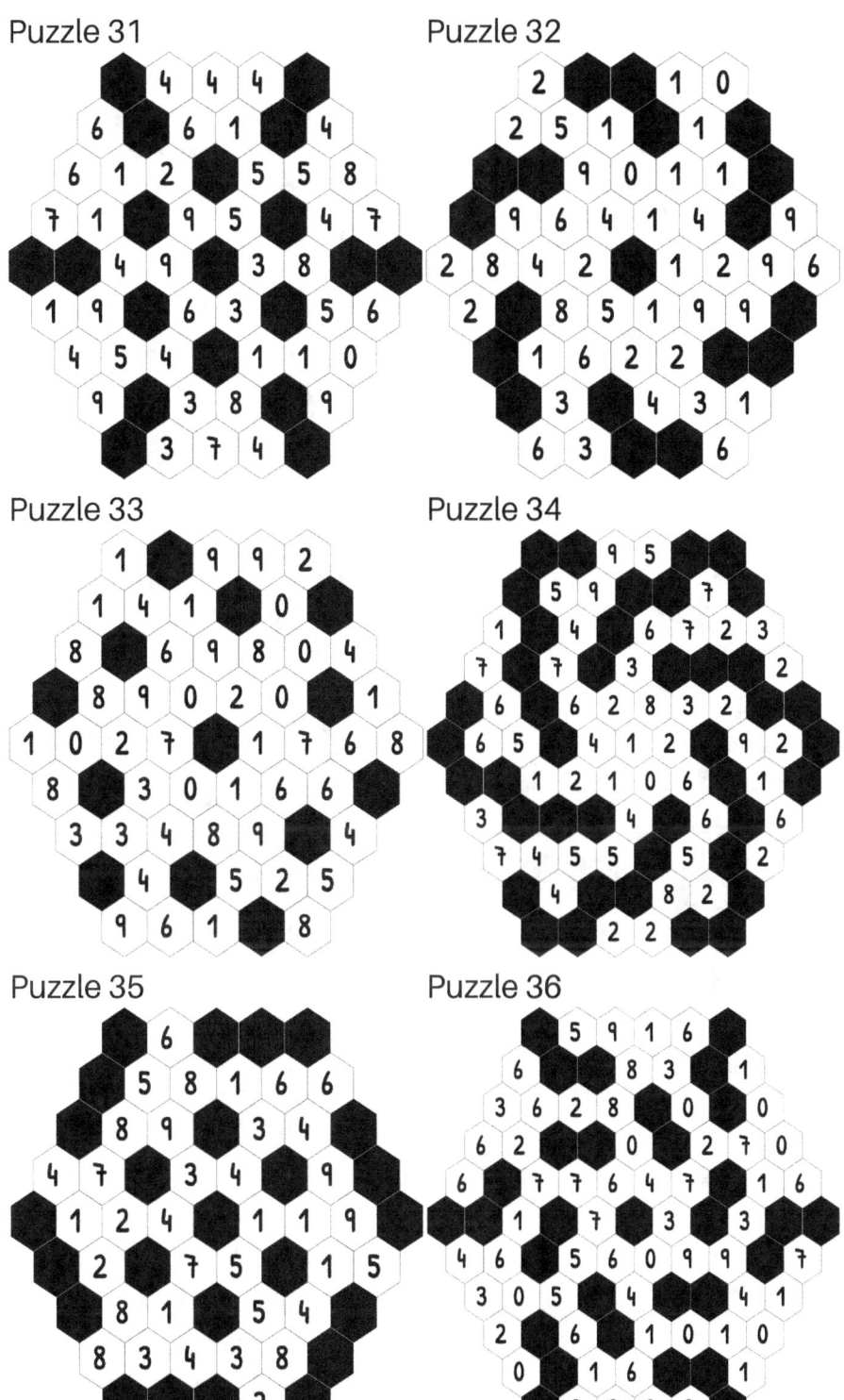

Puzzle 37

Puzzle 38

Puzzle 39

Puzzle 40

Puzzle 41

Puzzle 42

Puzzle 43

Puzzle 44

Puzzle 45

Puzzle 46

Puzzle 47

Puzzle 48

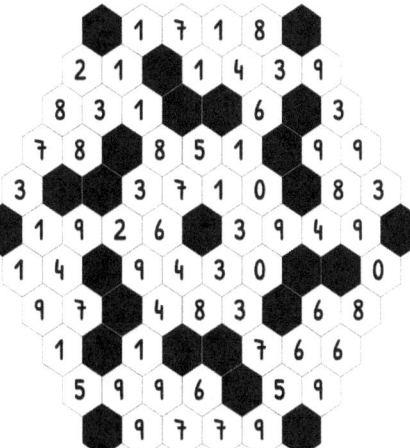

About the Author

Neil Aggarwal loves puzzles. Sometimes he is solving an existing puzzle, sometimes he is figuring out how to create new ones. Computer programming is always a good source of puzzles. So is using a new ingredient or making a new recipe. Sometimes the puzzle is to figure out how to get his 4x4 vehicle across a challenging trail. Life is full of puzzles, and he enjoys all of them. He also loves spending time with his family, friends, and dogs -- who, at times, are puzzles too!